Translational Bioinformatics Applications in Healthcare

Intelligent Signal Processing and Data Analysis

Series Editor:
Nilanjan Dey

Intelligent signal processing (ISP) methods are progressively swapping the conventional analog signal processing techniques in several domains, such as speech analysis and processing, biomedical signal analysis, radar and sonar signal processing, telecommunications, and geophysical signal processing. The main focus of this book series is to find out the new trends and techniques in intelligent signal processing and data analysis leading to scientific breakthroughs in applied applications. Artificial fuzzy logic, deep learning, optimization algorithms, and neural networks are the main themes.

Bio-Inspired Algorithms in PID Controller Optimization
Jagatheesan Kallannan, Anand Baskaran, Nilanjan Dey, Amira S. Ashour

A Beginner's Guide to Image Preprocessing Techniques
Jyotismita Chaki, Nilanjan Dey

Digital Image Watermarking: Theoretical and Computational Advances
Surekha Borra, Rohit Thanki, Nilanjan Dey

A Beginner's Guide to Image Shape Feature Extraction Techniques
Jyotismita Chaki, Nilanjan Dey

Coefficient of Variation and Machine Learning Applications
K. Hima Bindu, Raghava Morusupalli, Nilanjan Dey, C. Raghavendra Rao

Data Analytics for Coronavirus Disease (COVID-19) Outbreak
Gitanjali Rahul Shinde, Asmita Balasaheb Kalamkar, Parikshit Narendra Mahalle, Nilanjan Dey

A Beginner's Guide to Multi-Level Image Thresholding
Venkatesan Rajinikanth, Nadaradjane Sri Madhava Raja, Nilanjan Dey

Hybrid Image Processing Methods for Medical Image Examination
Venkatesan Rajinikanth, E. Priya, Hong Lin, Fuhua (Oscar) Lin

Translational Bioinformatics Applications in Healthcare
Khalid Raza and Nilanjan Dey

For more information about this series, please visit:https://www.routledge.com/Intelligent-Signal-Processing-and-Data-Analysis/book-series/INSPDA

Translational Bioinformatics Applications in Healthcare

Edited by
Khalid Raza and Nilanjan Dey

CRC Press is an imprint of the
Taylor & Francis Group, an **informa** business

First edition published 2021
by CRC Press
6000 Broken Sound Parkway NW, Suite 300, Boca Raton, FL 33487-2742

and by CRC Press
2 Park Square, Milton Park,Abingdon,Oxon,OX14 4RN

Library of Congress Cataloging-in-Publication Data
Names: Raza, Khalid, editor. | Dey, Nilanjan, editor.
Title: Translational bioinformatics applications in healthcare / edited by Khalid Raza and Nilanjan Dey.
Description: First edition. | Boca Raton, FL: CRC Press, 2021. |
Series: Intelligent signal processing and data analysis | Includes bibliographical references and index. |
Summary: "Translational bioinformatics (TBI) involves development of storage, analytics and advanced computational methods to harvest knowledge from voluminous biomedical and genomic data into 4P healthcare (proactive, predictive, preventive and participatory). This book offers a detailed overview and concepts of TBI, biological and clinical databases, clinical informatics, and pertinent real-case applications. It further illustrates recent advancements, tools, techniques, and applications of TBI in healthcare including IoT potential, toxin databases, medical image analysis and telemedicine applications, analytics of COVID-19 CT-images, viroinformatics and viral diseases, COVID-19 related research" — Provided by publisher.
Identifiers: LCCN 2020049426 (print) | LCCN 2020049427 (ebook) | ISBN 9780367705701 (hardback) | ISBN 9781003146988 (ebook)
Subjects: LCSH: Bioinformatics. | Drug development. | Medical informatics.
Classification: LCC QH324.2 .T734 2021 (print) | LCC QH324.2 (ebook) | DDC 610.285—dc23 LC record available at https://lccn.loc.gov/2020049426 LC ebook record available at https://lccn.loc.gov/2020049427

ISBN: 978-0-367-70570-1 (hbk)
ISBN: 978-1-000-37523-7 (pbk)
ISBN: 978-1-003-14698-8 (ebk)

Typeset in Times LT Std
by codeMantra

Contents

Preface

Translational bioinformatics is an emerging interdisciplinary field of study involving the development of storage, analytics, and advance computational methods to harvest knowledge from voluminous biomedical and genomic data into 4P healthcare (proactive, predictive, preventive, and participatory). Modern biology relies on high-throughput measurement techniques including next-generation sequencing (NGS), microarray technologies, proteomics, imaging technologies, and meta-genomics that have flooded massive amount of biological data giving rise to a number of biological databases. These biological and health-related databases are valuable resources and tools to various stakeholders, including biologists, clinicians, bioinformaticians, and healthcare professionals. The final outcome of translation bioinformatics is novel knowledge that can be better utilized in diagnosis, prognosis, and treatment of various diseases. *Translational Bioinformatics Applications in Healthcare* offers a detailed overview on concepts of translational bioinformatics, various biological and clinical databases, clinical informatics, and real-case applications of translational bioinformatics in healthcare. This book has a total of ten chapters that are logically grouped into three parts. We hope you will enjoy reading it. The aim of editing this volume is to bring together recent advancement, tools, techniques, and applications of translational bioinformatics in healthcare. This volume will be a valuable resource for health educators, clinicians, healthcare professionals, and graduate students of biology, biostatistics, biomedical sciences, bioinformatics, and interdisciplinary sciences.

Editors

Khalid Raza, PhD, is an Assistant Professor at the Department of Computer Science, Jamia Millia Islamia (Central University), New Delhi. Dr. Raza has been honored with "ICCR Chair Visiting Professor" by the Indian Council for Cultural Relations (ICCR), Ministry of Foreign Affairs, Government of India, and deputed at the Faculty of Computer & Information Sciences, Ain Shams University, Cairo, Egypt. He has more than 10 years of teaching and research experience in the field of translational bioinformatics and computational intelligence applications in healthcare. He has contributed more than 70 research articles in reputed journals and edited books. He has authored/edited books with reputed publishers. He has reviewed more than 150 research articles for reputed journals/conferences since 2015. Dr. Raza has delivered several keynote addresses, invited talks, public lectures, and seminars in national and international conferences, workshops, and chaired technical sessions at various conferences. He has served as the technical program committee member of several international conferences and workshops. He has also executed two Indian government-funded research projects. Dr. Raza is a member of MIR Labs (United States), CSI (India), and SCRS (India). His research interests include translational bioinformatics, computational intelligence methods and its applications in bioinformatics, viroinformatics, and health informatics.

Nilanjan Dey, PhD, is an Associate Professor, Department of Computer Science and Engineering, JIS University, Kolkata, India. He is a visiting fellow at the University of Reading, U.K. He was an honorary Visiting Scientist at Global Biomedical Technologies Inc., California (2012–2015). He was awarded his PhD from Jadavpur University in 2015. He has authored/edited more than 70 books with Elsevier, Wiley, CRC Press, and Springer, and has published more than 300 papers. He is the Editor-in-Chief of International *Journal of Ambient Computing and Intelligence*, IGI Global, and Associated Editor of *IEEE Access* and *International Journal of Information Technology*, Springer. He is the Series Co-Editor of *Springer Tracts in Nature-Inspired Computing*, Springer, Series Co-Editor of *Advances in Ubiquitous Sensing Applications for Healthcare*, Elsevier, and Series Editor of *Computational Intelligence in Engineering Problem Solving* and *Intelligent Signal Processing and Data Analysis*, CRC. His main research interests include medical imaging, machine learning, computer-aided diagnosis, and data mining. He is the Indian Ambassador of International Federation for Information Processing–Young ICT Group and Senior member of IEEE.

Contributors

Shaban Ahmad
ICAR-Agricultural Knowledge Management Unit
Indian Agricultural Research Institute
New Delhi, India

Rangel Arthur
Telecommunications Engineering
School of Technology (FT)
State University of Campinas (UNICAMP)
Campinas, Brazil

Punuri Jayasekhar Babu
Department of Biotechnology
Pachhunga University College
Mizoram University
Aizawl, India

Sushmita Baishnab
Centre for Biotechnology and Bioinformatics
Dibrugarh University
Assam, India

Urmi Bajpai
Department of Biomedical Science, Acharya Narendra Dev College
University of Delhi
New Delhi, India

Ana Carolina Borges Monteiro
Communications Department (DECOM)
School of Electrical and Computer Engineering (FEEC)
State University of Campinas (UNICAMP)
Campinas, Brazil

Mohinikanti Das
Department of Botany
College of Basic Science & Humanities
Odisha University of Agriculture and Technology
Bhubaneswar, India

Ravins Dohare
Centre for Interdisciplinary Research in Basic Sciences
Jamia Millia Islamia
New Delhi, India

Reinaldo Padilha França
Communications Department (DECOM)
School of Electrical and Computer Engineering (FEEC)
State University of Campinas (UNICAMP)
Campinas, Brazil

Lenin Fred A.
Department of CSE
Mar Ephraem College of Engineering and Technology
Elavuvilai, India

Arabinda Ghosh
Microbiology Division
Department of Botany
Gauhati University
Assam, India

Balázs Gulyás
Cognitive Neuroimaging Centre
Nanyang Technological University
Singapore

Agnik Haldar
Department of Bioinformatics
Central University of South Bihar
Bihar, India

Yuzo Iano
Communications Department (DECOM)
School of Electrical and Computer Engineering (FEEC)
State University of Campinas (UNICAMP)
Campinas, Brazil

Surabhi Johari
School of Biosciences
Institute of Management Studies University Courses
Ghaziabad, India

Fatima Nazish Khan
Department of Computer Science
Jamia Millia Islamia
New Delhi,India

Narender Kumar
Department of Mathematics
Gargi College
University of Delhi
New Delhi, India
Centre for Interdisciplinary Research in Basic Sciences
Jamia Millia Islamia
New Delhi, India

S. N. Kumar
Department of EEE
Amal Jyothi College of Engineering
Kerala, India

Ajay Kumar H.
Department of ECE
Mar Ephraem College of Engineering and Technology
Elavuvilai,India

L. R. Jonisha Miriam
Department of ECE
Mar Ephraem College of Engineering and Technology
Elavuvilai, India

Rina Ningthoujam
Department of Vegetable Science
College of Horticulture and Forestry
Central Agricultural University
Imphal, India

Parasuraman Padmanabhan
Cognitive Neuroimaging Centre
Nanyang Technological University
Singapore

Manasa Kumar Panda
Environment & Sustainability Department
CSIR-Institute of Minerals and Materials Technology
Bhubaneswar, India

Srinivasan Ramachandran
G. N. Ramachandran Knowledge of Centre
Council of Scientific and Industrial Research—Institute of Genomics and Integrative Biology (CSIR-IGIB)
New Delhi, India

Jyoti Rani
Department of Biomedical Science, Acharya Narendra Dev College
University of Delhi
New Delhi, India

Khalid Raza
Department of Computer Science
Jamia Millia Islamia
New Delhi, India

Shweta Sankhwar
Department of Information Technology
Babasaheb Bhimrao Ambedkar University
Lucknow, India

Mrinal Kumar Sarma
Advance biofuel Division
The Energy and Resources Institute
New Delhi, India

Ashwani Sharma
International Computational Center
Rennes, France

Kayenat Sheikh
Department of Computer Science
Jamia Millia Islamia
New Delhi, India

Yengkhom Disco Singh
Department of Post Harvest Technology
College of Horticulture and Forestry
Central Agricultural University
Pasighat, India

Ajay Kumar Singh
Department of Bioinformatics
Central University of South Bihar
Bihar,India

Subrata Sinha
Centre for Biotechnology and Bioinformatics
Dibrugarh University
Assam, India

Ankit Srivastava
Department of Opthalmology
Institute of Medical Sciences
Banaras Hindu University
Varanasi, India

Part I

Translational Healthcare, Next-Generation Sequence Analysis, and Drug Repurposing

1 Translational Healthcare System through Bioinformatics

Mrinal Kumar Sarma
The Energy and Resources Institute

Rina Ningthoujam
Central Agricultural University

Manasa Kumar Panda
CSIR-Institute of Minerals and Materials Technology

Punuri Jayasekhar Babu
Mizoram University

Ankit Srivastava
Banaras Hindu University

Maohinikanti Das
Odisha University of Agriculture and Technology

Yengkhom Disco Singh
Central Agricultural University

CONTENTS

1.1 INTRODUCTION

The field of translational bioinformatics is relatively young and fast, capturing the interest of academic circles as an important discipline of personalized healthcare and precision medicine. Advanced biological methods and technologies of analysis, along with interpretation, have opened up a new realm of exploratory endeavor. Microscopy as an invention and as a field of scientific discourse, commonly known as optics and imaging, allowed doctors and researchers to witness changes at the cellular level(Kalinin et al., 2016). Introduction of X-ray followed by magnetic resonance and other similar imaging technologies enabled monitoring of tissues and organs which was never possible before (Yu et al., 2018). Another recent addition to the above solution is "Big data" that translates the purpose of the application albeit the size. The term relates to the hypothesis-free approach of underlying experimental designs. Big data analysis is hypothesis generating rather than hypotheses driven (Golub, 2010). As we observe, innumerable data sets are generated when exploring a complicated disease worked on by researchers worldwide. These data points serve as feed for computational tools that process this information to get an apprehensible outcome which is used to target a specific medical condition. This approach of complementing complex experimental data volume with bioinformatics tools to arrive at a targeted solution for the health exigency is termed as translational healthcare (TH) system. This complex set of points also termed as "Big data" addresses the interconnectivity of different pathways at our biological levels, thereby probing the dysfunctional part of the body. The goal of precision medicine is to make use of this big data results and convert them into the information of pragmatic value for the practicing clinician.

The concept of TH is loosely based on the collection of big data provided by individual customers willingly and delivered directly to commercial giants like Facebook and Amazon, which helps them to tailor products according to their personal likes and dislikes. This approach has been thought over in medical and healthcare domain as a collection of information about a patient's clinical history that can be put together and assessed using a computational tool, resulting in a possible personalized, specific solution addressing the medical conditions to improve outcomes (Manyika et al., 2011). Compared to the commercial industry, healthcare sector is more stringent as information available to researchers are not easy due to statutory protection of patient's medical data. The direct inaccessibility to medical data by big data professionals, which is both philosophical and practical, is also a huge barrier in TH approach. Medical research in their field of research is still anchored in its classical approach of producing knowledge through studies and narrowing down the research question avoiding complexities in the real-world practice. Mostly, these kinds of studies are poorly equipped to correlate the interconnecting factors associated with the outcome of the treatment administered to the patient. The medical data generated in the hospitals on a daily basis can be a huge fuel for knowledge to be gained in treating an ailment. However, despite having so much information to develop an improved process for medical methods, big data analytics till date is limited only to the field of bioinformatics. The workers involved in this vibrant field encompass these details and develop user-friendly computational methods to deliver outcomes in healthcare that are understood by the practitioner. This new insight in the field of healthcare allows us to ideally treat disease and also in other case provides insight in preventing it in the first place.

1.2 DATA AND BIOMEDICINE

With the advent of modern technology like next-generation sequencing (NGS) and advanced molecular biology tools for genomics and transcriptomics, a comprehensive look at the experimental outcomes is increasingly becoming multidimensional. This new facet of pursuit requires a large number of feed points to present a true signal corroborated by the statistical analysis (Kulynych and Greely, 2014). The last decade has seen an ever-increasing number of large-scale biorepositories targeted toward clinical and TH research all over the planet. These biorepositories contain both soft information and biospecimens collected from patients, enabling researchers to define and design treatment based on the understanding of the disease by reclassifying it on the basis of their underlying metabolic pathways with the help of the bioinformatics tools instead of the classical approach toward clinical treatment that has been practiced for centuries (Denny, 2014). These projects involve data collection using various modes of operation, starting from explicit information provided by patients expressing consent to their medical history being used to de-identified specimens whose clinical records are obtained from EHRs (electronic health records, also de-identified). This model of data collection using ethically informed consent is rigorous and expensive, whereas the model of using the information of de-identified specimens is more scalable and financially viable. However, with the increasing use of genomic data, nowadays, it is very complicated and, in a way, impossible to de-identify these information sources (Fleurence et al., 2013). This has led to the increase in ethical issues concerning patient privacy and data sharing. Most of the developed countries have come up with legislation to address these issues (Hudson and Collins, 2015). To augment toward the process of learning healthcare systems (LHS) model, the partnership between the patients and their doctors or researchers is sought by empowering the patient as a collaborator of the project, who also has a part in it. TH practice is a transition from best evidence accrued via randomized clinical trials to practice-based evidence, i.e., application of data generated from real world rather than from controlled clinical trials or experiments. In fact, currently used experimental design of randomized clinical trials is expensive and at times tends to be different from practical scenarios mainly due to non inclusion of common comorbidities associated with the addressed medical question (Luce et al., 2009). The data collected are statistical interpretations in nature where there is an effort to provide a generalized solution to a disease, not taking into account the complexities arising in the lower average groups of the cohort studied. These cohorts are meant to be a representative of the population in consideration but in reality are farfetched from actualities. In LHS or biomedicine, the operation is bidirectional where research is used to inform practice, and the data points collected during the disease treatment can be applied for hypothesis generation along with its validation through pragmatic experiments. Data derived from the outcomes of these experiments, accessed through computational "Big Data" methods, can possibly provide guidelines for future clinical care and practice.

1.3 GENOMICS AND BIOINFORMATICS

The study of the genome of an organism, including whole DNA sequencing methods (Figure 1.1), gene mapping, and analyzing of intragenomic phenomena, is called genomics. It helps in pointing out the ideal gene or genotype among others. It has been globally used in healthcare and agriculture. In agriculture, it is used as a tool for programmed crop improvement. With the help of this process, we can generate new hybrid strains with good nutritional quality and more stress tolerance. The

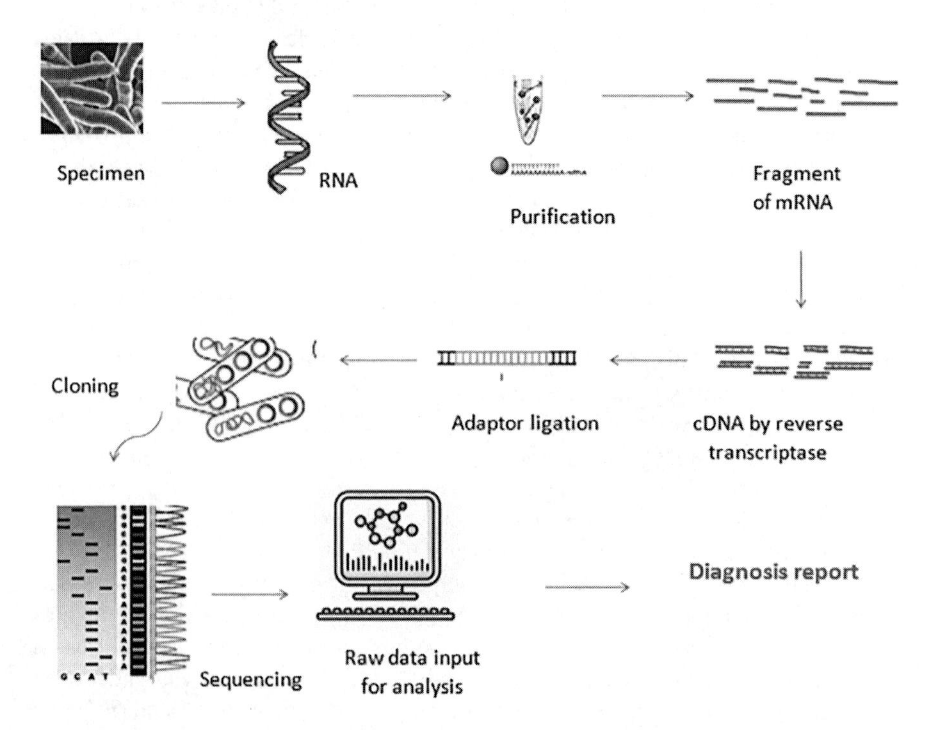

FIGURE 1.1 The stepwise process of genome sequencing to an analysis by bioinformatics.

potential impact of this technology in the clinical application includes the discovery of gene and diagnosis of genetic abnormalities; identification and treatment of common diseases such as high blood pressure, diabetes, and cancer; targeted therapy; noninvasive prenatal testing; and identification of human infectious microorganism genome. It is further used in gene therapy and editing. Likewise, bioinformatics is also a tool which is capable of handling an enormous amount of data that result from programmed genomics and proteomics. It is used in other scientific fields and helps in analyzing transcription and functional structure determination (Zagursky and Russel, 2001). It has the ability of data acquisition, storage, processing, analysis, and integration. This tool functions in three main steps (Luscombe et al., 2001). First, it organizes the data in the simplest form and gives the existing information to the researchers. Second, the data are analyzed by developing a tool and resource. Third, the results are provided in a biological manner.

To complete all these steps, different well-known databases are used. Some of them are Genetic Data Bank (GenBank), Protein Data Bank, SWISS-PROT, Protein Information Resource, Familial and Structural Protein Relationships, and Hierarchical Classification of Protein Domain Structures. By using these databases, basic research and modeling can be done with the help of sequence analysis tools like Basic Local Alignment Search Tool, Fast-All, CLUSTALW, etc. The visualization of these modeled structures is again done by using tools like WebLab, MOLMOL, Rasmol, etc. The benefit of this new technology has been used in different disciplines and is accepted globally. The advance method of this technology has advantage over other methods and has scope in the pharmaceutical industry and drug discovery, forensic analysis, crop improvement, food analysis, healthcare, environment, and biodiversity management. In humankind, the use of such type of technology may increase the rate of curing of inherited human diseases by developing new medicines. Currently, genomics and bioinformatics play an important role in investigating infectious diseases caused by various microorganisms to humankind and in the treatment of human genetic disorders.

Both genomics and bioinformatics are correlated with each other. Their contribution in the implication of healthcare has emerged slowly. The accumulated data of genomic and biomedical researches are managed and analyzed for healthcare with the help of bioinformatics (Nagarajan, 2004). The application of this science technology has a direct impact on understanding the mechanism of infectious diseases, the relation between pathogen and host, and transmission cycles (Degrave et al., 2002). Bioinformatics can be used in the classification of disease according to their genetic structures (Brown et al., 1999). They help in managing the highly infectious diseases caused by bacterial pathogens such as *Mycobacterium tuberculosis* and parasites such as *Plasmodium falciparum*. *Mycobacterium tuberculosis* causes tuberculosis, which leads to loss of lives in about two million people per year globally (WHO, 2013). The genome sequencing of the members of mycobacteria was first done in 1998 (Cole et al.,1998; Yun et al.,2016), and it helps in finding character of gene, development of new diagnostic equipment, drug susceptibility testing, and molecular epidemiology of circulating mycobacterium stains. The genome sequencing of other deadly diseases such as dengue fever, malaria, and filariasis is also done with this tool. Genomics and bioinformatics monitored the outbreak of disease, screening variation in the genome, mechanism of host immune evasion to identify the useful diagnostic markers and vaccine targets. They first generate the genomic data of the pathogen and then genomic of the host bringing together an animal model. Then, the health report of the patient is recorded to initiate the treatment options: drugs, vaccine, or a particular therapy (Srinivasa et al., 2009). The importance of genomics, proteomics, and bioinformatics is not only in the application of healthcare to humankind but also used in identifying traits gene of animals. They can be utilized for modification of transgenics to improve growth rate and carcass composition,increase feed utilization, improve changes in milk composition, increase mohair production,increase reproductive performance, and generate more resistivity to diseases (Baldassarre et al., 2004). The tools of bioinformatics are also applied in the screening of new castle diseases and bringing of new molecular diagnostics (Soetan and Abatan, 2008). Hence, the advancement of these innovative tools brings beneficial changes in healthcare framework and biomedical industry.

1.4 PHARMACOGENOMICS

Pharmacogenomics (PG) (pharmaco=drug; genomics=study of gene) is a branch of science that deals with how human gene has an impact or response to drugs. This field has evolved to improve the efficacy and reduce the side effects of administered drugs. Mostly, the majority of the drugs administered to the patient fulfil the cause and are also responsible for some adverse side effects. When a drug is administered to a patient, it has to reach its target through the bloodstream, act on the target, and finally come out of the body. After the targeted point is reached, the drugs are then finally absorbed, distributed, metabolized, and eliminated by the body. This whole process is eased by pharmacokinetic (PK) genes. These genes affect the ADME process (absorption, distribution, metabolization, and excretion) of a drug. On the other hand, the action of a drug on its target is facilitated by the pharmacodynamic (PD) genes which are mainly responsible for the desired clinical outcome. These PK and PD genes are involved in two different processes called intentional (on-target) and unintentional (off-target) (Karczewskietal., 2012). The intentional process involves desired therapeutic outcome, whereas unintentional process results in causing adverse events. Therefore, genetics plays a critical role in both adverse and desired events in a living organism, especially in determining optimal drug dose for an individual. For example, warfarin and clopidogrel, anticoagulant drugs, have different therapeutic doses based on the genetic makeup of an individual (Karczewskietal., 2012). Currently, research community is working to discover the genes engaged in PK and PD pathway genes. This could help in understanding the effect of drug action and to improve dosing. The overview of PG is illustrated in Figure 1.1.

In 1950s, a child suffering with leukemia was treated with mercaptopurine (a drug used as myelosuppressant), and the child began experiencing unexpected immunosuppression and bone marrow toxicity. However, later in 1990s researchers realized that genetics could explain the reason for the life threatening bone marrow toxicity (Abbott, 2003). Succinylcholine, a drug that came into market in 1950sand was used as a muscle relaxant by anesthesiologists during operations, was found to cause horrific respiratory arrest in about 1 in 2,500 individuals. Thus, there is an increasing need to understand the interaction between human genetics and drugs to formulate modern and personalized medicine.

1.5 MECHANISM OF PG

The genomic factors influence the two main interactions between drugs and body known as pharmacokinetics and pharmacodynamics (Figure 1.2).

1.5.1 Pharmacokinetics

When a drug is administered to a patient, it interacts with many proteins on the way to its target inside the body and on its way out of the body. These interactions determine the drug pharmacokinetics: how it gets absorbed, distributed, metabolized, and excreted. One of the important parameters of a drug is bioavailability (amount of drug in systemic circulation), which depends on the mode of its administration into a body. The drugs which are administered intravenously may 100% be available in the systemic circulation. On the contrary, drugs administered orally show very less availability in systemic circulation due to the influence of gastric emptying (i.e., transit time) and enzymatic action of intestine. For example, polymorphism in the class of *ABC* (ATP binding cassette) genes is associated with altered bioavailability

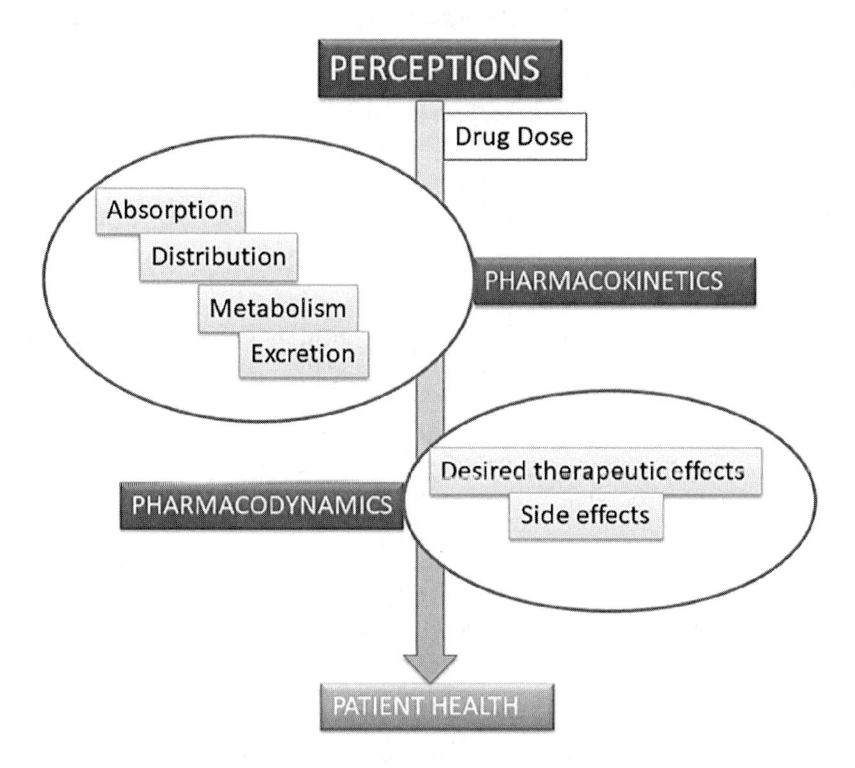

FIGURE 1.2 Overview of pharmacogenomics.

of certain cardiac drug digoxin drugs (Dietrich et al., 2003). However, the efficacy of a drug is determined by an individual's genomics rather than its mode of action inside the body. For example, cytokine CYP2D6 (a protein) is responsible for metabolism of many drugs. This *CYP2D6* gene variation determines whether an individual is a fast or slow metabolizer to a particular drug. If the patient's *CYP2D6* gene shows any variation to a particular drug, then they are considered as a fast metabolizer. In fast metabolizer's body, the drug shows very less effectiveness as it is broken down quickly and eliminated from the body immediately. In case of a slow metabolizer, there is no significant variation in patient's *CYP2D6* gene; however, they are susceptible to side effects because the drug concentration builds up in the body (Gopisankar, 2017).

1.5.2 Pharmacodynamics

Pharmacodynamics is concerned with both the therapeutic outcome and side effects of a drug which are greatly influenced by the genomic factors. Notably, some gene variants are linked with a serious adverse reaction. Example, carbamazepine is a popular and highly effective epilepsy drug. On administration, this drug induces the gene variant known as HLA-B*15:02, which increases the risk of severe hypersensitivity reactions. In other cases, some drugs may resemblance with the substrate of the proteins (enzymes). One such example is warfarin, an anticoagulant which binds to the catalytic site of the VKORC1 (vitamin K epoxide reductase complex subunit 1). The VKORC1 is an enzyme that typically converts the inactive form of vitamin K to active reduced forms (Zimmermann and Matschiner, 1974). Warfarin inhibits the reduction process (to form active form of vitamin K) ensuing lack of active vitamin K. This phenomenon of warfarin is eventually responsible for downstream regulation of active vitamin K formation (Oldenburg et al., 2007; Owen et al., 2008). Therefore, polymorphisms of VKORC1enzyme give a better idea of optimal dose of warfarin.

1.6 PG IN DRUG DISCOVERY

PG plays a key role in drug discovery, especially in identifying a possible gene target for a specific disease, minimizing cost, increasing the safety, and narrowing down the number of targets to be analyzed biochemically. Importantly, PG determines the potential polypharmacological factors that could contribute to adverse events. Also, it helps in dosing and efficacy of a particular drug after initial trails. There are three key steps involved in drug discovery.

(i) *Small Molecule Candidate Identification*: Finding a suitable gene to target is a key starting point in developing a drug for a disease. In general, the gene involved in the disease can be identified by exome sequencing, analysis of RNA expression profiles, and other biochemical analyses. Once the potential pathway or targets are identified, cheminformatics methods can be used to generate the predictions for potential "leads" (drugs). Once the predictions are generated, they have to be confirmed by biochemical methods.

(ii) *Clinical Trial Pipeline*: If the drugs are identified as potential molecules in biochemical methods, these drugs proceed with animal experiments for toxicity verification. The drug efficacy and safety are demonstrated during the Phase II and carried even in Phase III. Sometimes patient response to a drug is different from the initial Phase II trials. This response is often related to PK and PD genetic factors suggesting that the PG can be used to limit the cohort for Phase III trials (Roses, 2004).

(iii) *Drug Repurposing*: As mentioned earlier, cheminformatics methods can be used to predict and identify a novel drug–protein interaction which helps in discovering new small molecules (drugs) for therapeutic application. However, drug "repurposing" is a strategy for identifying new uses of a certain drug that is already approved by FDA for a certain medical indication, for example, entacapone, a drug discovered and approved to treat Parkinson's by inhibiting catchol-O-methyltransferase. The docking studies predicted that entacapone binds to InhA (the enoyl-acyl carrier protein reductase). Based on these predications, biochemical evaluation of entacapone is performed, and it is shown to have anti-*Mycobacterium tuberculosis* activity (Kinnings et al., 2009). Thus, it is clearly understood that the pharmacological genomics plays an important role in drug discovery, the efficacy of a drug, and, more importantly, minimizing the side effects of a drug.

1.7 DRUG DISCOVERY AND DEVELOPMENT THROUGH OMICS TECHNOLOGIES

1.7.1 Omics Technology

It is well known that the bioinformatics approach is used widely for the identification of genes, proteins, and pathways indulge in diseases that could be targeted as therapeutics. System biology and computational method along with omics technology refer to the study of mechanism, interaction, and function of cellular molecules at the level of genome, transcript, protein, and metabolites(Keusch, 2006). Moreover, omics technique uses high-throughput data because they investigate huge number of genes, protein, and other products in targeted and nontargeted manner. The data generated by high-throughput studies generally require computer-based tools for prediction and analysis. With the advancement in disease progression and change in symptoms day by day, it is important to understand the molecular/genetic analysis of the diseases. Therefore, the detection and treatment of the disease is greatly felicitated by the 'omics' technology including genomics, proteomics, metabolomics, cytomics, and bioinformatics (Figure 1.3). In the present scenario, bioinformatics and omics together become emerging tools which are used not only for the study of early detection of cancer, cellular metabolism, cell growth and death, the discovery of biomarker but also for the prognosis and treatment of diseases by elucidating the mechanism of action of various drugs, interacting network in searching the drug target, identifying the potential target, and therapeutic drugs and their side effects.

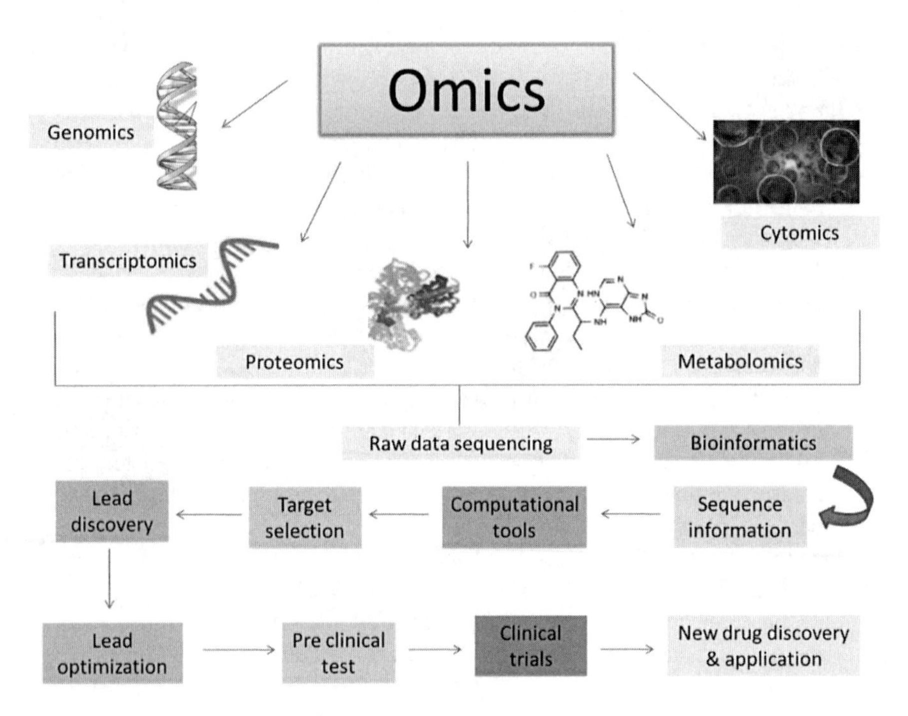

FIGURE 1.3 Omics-based technologies in drug discovery and development through computational tools.

1.8 AVAILABILITY OF OMICS TECHNOLOGY IN DRUG DISCOVERY AND DEVELOPMENT

In the era of TH system, omics technology is considered as the pillar of translational medicine and research in drug discovery and development including genomics, transcriptomics, proteomics, metabolomics, cytomics, and bioinformatics. The rise of translational medicine and the availability of data generated through high-throughput technology are used to develop clinical care and practices and also to manage and explore these data, which is the key responsibility of the translational researchers.

1.9 ROLE OF GENOMICS IN DRUG DISCOVERY AND DEVELOPMENT

Genomics is the study of the whole genome of an organism encoded in the form of nucleic acid (DNA or RNA), including the coding and noncoding region (Min Jou et al., 1972). Genomics applications are not only limited to discover the genes and its impact on tumorigenesis in cancer patients (Shih and Wang, 2005),altered molecular pathways associated with the disease, identification of biomarker but also applied for the treatment of disease using various drugs, validation of the existing drug response during the diseased condition. Drug discovery is a time-consuming process (Kola and Landis, 2004), and its rate of failure and fidelity could be enhanced by applying genomic technology. Using the genomic research, one can improve the drug discovery and development by change in DNA sequence such as DNA rearrangement, DNA copy number, single nucleotide polymorphism, mRNA profiling, use of genomic biomarker, DNA methylation, developing new target for molecular pathways, and microarray technology.

1.10 ROLE OF TRANSCRIPTOMICS IN DRUG DISCOVERY AND DEVELOPMENT

Transcriptomics provides lineage between genomics and proteomics. Transcriptome study entails the expression of genes at transcript (mRNA) level, and their expression is correlated with biological function. Variety of techniques have been considered under transcriptomics including suppression substrate hybridization, serial analysis of gene expression, RNA arbitrarily primer–polymerase chain reaction,total gene expression analysis (TOGA),and quantitative reverse transcription polymerase chain reaction. Currently,the most accepted tool is "DNA microarray," which enables to measure the expression of thousands of genes simultaneously and benefits all phases of drug discovery and validation. Gene expression profiling through microarray can be used to analyze cell cycle, signal transduction, the pharmacological effect of compounds, identification and validation of potential biomarker and new molecular targets (Zhang, 2007), action mechanism of a drug, drug resistance, sensitivity, and toxicity. All these major processes become a promising approach for drug discovery and development.

1.11 ROLE OF PROTEOMICS IN DRUG DISCOVERY AND DEVELOPMENT

With the advancement in translational research in medicine, proteomics approach is becoming an important tool to study the structural, functional, and regulatory roles in the cell. In addition, various computational techniques are also available for proteome analysis of biological samples in pathological conditions such as protein microarray, 2-D electrophoresis, mass spectrometry, and yeast two-hybrid system. Protein profiling with X-ray crystallography, nuclear magnetic resonance, and circular dichroism can be of high use in clinical diagnostics. With the era of translational and personalized medicine, the abovementioned proteomics technologies gain more advertisement to the field of new drug discovery and development to target and stop the progression of the disease by inactivating the protein involved in disease (Arora et al.,2005).

1.12 ARTIFICIAL INTELLIGENCE AND MACHINE LEARNING OF DRUGS

Artificial intelligence and machine learning scrutinize the different techniques and methods of empowered machines knowingly to achieve the optimal results (Mnih et al., 2015). It establishes machine learning for the advancement of design executions contingent on complicated structures of the impartial system and deep learning, which is distinguished in recent kinds of literature as the potential machine to improve or expand computerized platforms (Panteleev et al., 2018). Artificial intelligence was reawakened into society by Alpha Go in March 2016 (Jing et al., 2018). Nowadays, it is widely used in many fields and is gaining popularity for its ongoing performance in computer technology,machinery,telecommunication,robotics,autom ation,engineering,systemization,icon evaluation, language interpretation, and so on (Sellwood et al., 2018).

The basic intention of using artificial intelligence and machine learning is to invent or design the best quality drug to meet a certain therapeutic necessity (Butina, 1999). This method can be benefited more effectively by identification of proper hits from high-throughput screening followed by lead optimization which requires less time for drug designing and also decreases the price (Cheng et al., 2002), which includes proper identification of drug destination, special composition, upturn of patterns, matter conformations, preclinical experiments, and diagnostic analysis (Sellwood et al., 2018).

Machine learning techniques are generally considered to be conventional methods for the innovation and evolution of new drugs which are extensively being used for constructing prophetic patterns such as quantitative structure-activity and structure-property association patterns (Svetnik et al., 2003). Many chemists and biochemists have reported various drug designs that are successfully achieved by these methods with the enforcement of chemical and computative limitations (Blaney et al., 1982). Here the drug inventor sets up a new pattern for chemical structure by inserting functional groups at their appropriate locations (Goodford, 1985). The functional groups for drug designing can be determined by X-rays, probe evaluation, surface examinations, and the designer's expertness, experience, creativity, and analytic skills (Böhm, 1992).

Drug and disease are two fundamental components to regenerate a drug. On account of interrelationship with heterogeneity, network analysis is used to illustrate these components (LotfiShahreza et al., 2017). There are nine network systems that manipulate together for recycling one drug with better quality (Wang et al., 2014).

Artificial intelligence showing its wide application in research, medical, and biotechnological fields has the ability to reconstitute the drug discovery method, which has been proved by widespread usage of assimilation, absorption, dispersion, secretion, virulence appliances, and effective screening (Leelananda et al., 2016). Its application associates with statistics planning, collection of suitable info, categorization, organization, deterioration, and expansion (Duch et al., 2007). It is also used in the clinical drug development process for the enhancement of optical conception to detect clinically important samples like data visualization, recognition, and segmentation (Topol, 2019).

1.13 REAL TIME BIOMEDICAL AND HEALTHCARE DATA

Over the past decade, rapid development and advancement in the biomedical and healthcare informatics through NGS, translational bioinformatics,clinical and molecular imaging,sensor informatics, and most important omics (genomics, transcriptomics, proteomics, metabolomics etc.) generate large amount of data constitute "big data."

Availability of big data is successfully used by the bioinformatician, computational biologist, and information technologist for testing new hypothesis, data management and analysis for improvement of healthcare wellness,disease management,biomarker discovery for prognosis and diagnosis of disease, and clinical diagnosis. Adversely,the data generated through abovementioned technologies have also raised the problem in terms of security, data privacy, governance, and data ownership. Despitethe popularity of data exploration,integration of biomedical and healthcare is expected to increase with the help of technology upgrade in recent years.

1.14 BIOMEDICAL AND HEALTHCARE CHALLENGES

This new innovative technology challenges the release of effective treatments and control or preventive measures as the incorporation ofbasic science in medical care is considered to be sustained. Major challenges in biomedical and healthcare consist of storing,managing, and analyzing large amount of datasets (Margolis et al., 2014) and difficulty in management of EHRs. It is due to different clinical terminologies and specification, regular cleaning of data is required to ensure correctness, consistency and accuracy. Further, protection of healthcare data from phishing and hacking, up-to-date stored metadata, and immediate data sharing for patients convenience are some of the important requirements. Some other difficulties today are slow pace of technological advancement, inappropriate utilization of product, failure to promote the new product in a profitable way, and loss of mutual trust among patients, health workers, and researchers. The starting of a novel technology needs funding, cooperation among scientists, good relation with government, foundation, and universities, and a positive relationship with private and public companies. Experts with specialized technical skills are needed for work. Moreover, sharing of information on a result between researchers of different institutes and big industries may bring positive effects to work faster. Cost management for implementation of better healthcare and diagnosis is also a most important task as the research is expensive and capital intensive (Qazi and Raza, 2020).

1.15 OPPORTUNITIES TO IMPROVE BIOMEDICAL AND HEALTHCARE SECTORS

There is growing concern to improve the healthcare facilities with the emergence of technology upgrade, development of novel hardware and software for the computational purpose, NGS machines, and expansion of EHRs. Healthcare informatics is defined as a field to use of information, knowledge and data to improve public health, healthcare and research (Hersh, 2009). In addition, informatics plays an important role in driving and facilitating the science for the new discoveries in the datasets with better management and analysis. Through informatics, not only patients are benefited but also providers through the exchange of health information with more safety, security, and privacy. Further NGS, genome-wide association studies, and phenome-wide association studies are effectively providing information at depth using clinical data from EMR linked to the patient's sample and also record biological data (Raza, 2017; Raza and Ahmad, 2019) in a real-time manner. NGS technology has greatly simplified the sequencing technique with increased volume of biomedical and healthcare dataat reduced costs (Service, 2006). Era of "omics technology" also aids to provide enough amount of healthcare datasets in conjugation with systematic and integrative biology, which will allow us to understand the mechanism and progression of the complex disease such as cancer and development of disease-specific biomarker.

1.16 DATA MODELS FOR HEALTHCARE AND WELL CARE ANALYTICS

Healthcare data analytics is also known as clinical data analytics. It involves the collection of data that is relevant to healthcare such as claims and cost data, pharmaceutical data, research data, and clinical data that are collected from electronic medical records and patient's data. Many hospitals, including the private sector and medical institutes, cope with this data analytics. With the help of data analytics, better care about the patient can be performed, and with efficient use of data, it is applied in quantitative and qualitative analysis (Simpao et al., 2014). In developed countries like United States, healthcare analytics is expected to raise more than $31 billion by 2022 (Healthcare Finance news, 2018). According to the sources of data for healthcare system (Senthilkumar et al., 2018), it can be widely divided as structured data, semi-structured data, and unstructured data. Structured data is data that are highly organized and easily addressable for more operative processing and analysis. They are well structured, formatted repository, and typically a database—for example, well-organized data information of hierarchical terminologies of many diseases, signs and symptoms, preventive and treatment information, laboratory results, information about the ill person such as date of admission, name of uptake drugs, and billing histories in the hospital or clinical services. Semi-structural data refer to data that are organized without following the formal structure of data models. An example of this type of data source includes the generation of data from sensors of devices to result in an effective monitoring of patient's manner. Unstructured data files does not follow any specific rules and sequence. There is no predefined model, and they are not capable of being searched easily. They may be audio, video, clinical letters in word processing documents, etc. Healthcare data analytics also help in determining the improving techniques of clinical care. The utilization of analytics in healthcare, namely, data mining,text mining, and big data analytic, can predict the diagnosis and treatment of the patient. It leads to improvement in service quality and reduction of treatment cost. Among these various types of healthcare analytics, data mining is commonly applied in clinical decision support (42%). Data mining is observed utilizingthe treatment of coronary artery disease (Tsipouras et al., 2008). It is again found using existing data such as demographics,medical history,simple physical examination, blood test, and noninvasive simple investigations. Data mining is uncomplicated, low cost, and can get anaccurate result. Application of data mining is also done to screen out rare type of diabetes,factors of controlling diabetes, and recorded patient information to get knowledge. Different researchers get knowledge in diagnosis of diabetes and identify the unique and difficult form of diabetes by using the three-level clustering framework from recording the history of ill patients (Antonelli et al., 2013). Application of data mining in treatment of cancer helps to identity the condition of patient faster. The setting of this data makes it easier to know the suitable and matching treatment for patients. It is useful in screening of head injury, taking of CT scan, X-ray, body testing like blood glucose, etc. (Ceglowski et al., 2007). Other beneficial application of data mining includes prediction of ulcer formation, general problem lists, and personalized medical care. The experiment on data mining and big data analytics is reported in the past 10 years both in the form of theory and applications (Su et al., 2012). The analytics of clinical data need to follow a pattern of healthcare analytics adoption model. The framework of this system is developed after ensemble of industries as a way to differentiate the capability of analytics and result in a well-organized sequencing system. The adoption model of healthcare analytics can be represented in eight advanced levels. They are fragmented point solutions, enterprise data warehouse, standardized vocabulary and patient registries, automated internal reporting, automated external reporting, waste and care variability reduction, population health management and suggestive analytics, clinical risk intervention and predictive analytics, and personalized medicine and prescriptive analytics. The successful entry and organization of data lead to effective analytics. Currently, the outcomes from these analytics help in the management of population health. Therefore, it should be promoted more to result in a positive effect on healthcare.

1.17 CONCLUSION AND FUTURE PROSPECTS

Development and advancement in biomedical and healthcare sectors provide a lot of information in day-to-day life; thus, advances in information processing for bioinformatics, sensing, and imaging technologies will have a great impact on future clinical research. The collection of generated data through biomedical tools is continuously helping in building a better personalized healthcare and prognostic framework. To effectively manage, analyze, and interpret the data, healthcare workers needed to group together with a computational biologist, a data scientist, and engineers. In addition, there is an urgent need to educate healthcare practitioners on how to handle these challenges.

A careful integration of streams of biological data and their application in clinical setup is highly important in precision medicine. Big data analysis with improved technologies and predictive models in dealing with different diseases can be discussed for future prospect. However, individualized medicine is going to play major role in healthcare system. The future offers translational bioinformatics with new discovery, development, drug design, and new therapeutics applications.

Several multi-omics high-throughput technologies through computational biology, system biology, and bioinformatics allow us to understand the disease progression and treatment response, molecular basis of pathogenesis, application use in the drug development process, drug target and validation, mechanism of action of various drugs, toxicity and safety efficacy. Hence, improvements need attention to revolutionize the omics technologies in the TH system.

ACKNOWLEDGMENTS

The authors are thankful to their respective institution. YDS is very much thankful to the Vice Chancellor, Central Agricultural University, Imphal, for providing facilities.

REFERENCES

Abbott, A. With your genes? Take one of these, three times a day. *Nature* 425 (2003): 760–762.

Antonelli, D., Baralis, E., Bruno, G., Cerquitelli, T., Chiusano, S., and Mahoto, N.Analysis of diabetic patients through their examination history. *Expert Systems with Applications*, 40, no. 11 (2013): 4672–4678.

Arora, P.S., Yamagiwa, H., Srivastava, A., Bolander, M.E., and Sarkar, G. Comparative evaluation of two two-dimensional gel electrophoresis image analysis software applications using synovial fluids from patients with joint disease. *Journal of Orthopaedic Science* 10 (2005): 160–166.

Baldassarre, H., Wang, B., Keefer, C.L., Lazaris, A., and Karatzas, C.N., 2004. State of the art in the production of transgenic goats. *Reproduction, Fertility and Development* 16, no. 4 (2004): 465–470.

Blaney, J.M., Jorgensen, E.C.,Connolly, M.L.,Ferrin, T.E.,Langridge, R., Oatley, S.J., Burridge, J.M., and Blake, C.C.F. Computer graphics in drug design: molecular modeling of thyroid hormone-prealbumin interactions. *Journal of Medicinal Chemistry* 25, no. 7 (1982): 785–790.

Böhm, H.-J. The computer program LUDI: a new method for the de novo design of enzyme inhibitors. *Journal of Computer-Aided Molecular Design* 6, no. 1 (1992): 61–78.

Brown, N., Nelis, A., Rappert, B., Webster, A., and Ommen, G.J.B.V. 1999. Bioinformatics: A technology Assessment of Recent Developments in Bioinformatics and Related Areas of Research and Development Including High throughput Screening and Combinatorial Chemistry, *Final Report for the Science and Technological Options Assessment (STOA) Unit*, European Parliament.

Butina, D. Unsupervised data base clustering based on daylight's fingerprint and Tanimoto similarity: a fast and automated way to cluster small and large data sets. *Journal of Chemical Information and Computer Sciences* 39, no. 4 (1999): 747–750.

Ceglowski, R., Churilov, L., and Wasserthiel, J., 2007. Combining data mining and discrete event simulation for a value-added view of a hospital emergency department. *Journal of the Operational Research Society* 58, no. 2 (2007): 246–254.

Cheng, A., Diller, D.J., Dixon, S.L., Egan, W.J., Lauri, G., and Merz Jr, K.M. Computation of the physio-chemical properties and data mining of large molecular collections. *Journal of Computational Chemistry* 23, no. 1 (2002): 172–183.

Cole, S., Brosch, R., Parkhill, J., Garnier, T., Churcher, C., Harris, D., Gordon, S.V., Eiglmeier, K., Gas, S., Barry, C., and Tekaia, F. Deciphering the biology of Mycobacterium tuberculosis from the complete genome sequence. *Nature* 393, no. 6685 (1998): 537–544.

Dash, S., Shakyawar, S.K., Sharma, M., and Kaushik, S. Big data in healthcare: management, analysis and future prospects. *Journal of Big Data* 6, no. 1 (2019): 54.

Degrave, W., Huynh, C., Roos, D., Oduola, A., and Morel, C.M. Bioinformatics for disease endemic countries: opportunities and challenges in science and technology development for health. *IUPAC Journal* 1(2002): 1–7.

Denny, J.C. Surveying recent themes in translational bioinformatics: big data in EHRs, omics for drugs, and personal genomics. *Yearbook of Medical InformaticsYear b Med Inform* 9(2014): 199–205.

Dietrich, C.G., Geier, A., and Oude Elferink, R.P. ABC of oral bioavailability: transporters as gatekeepers in the gut. *Gut* 52 (2003): 1788–1795.

Duch, W., Swaminathan, K., and Meller, J. Artificial intelligence approaches for rational drug design and discovery. *Current Pharmaceutical Design* 13, no. 14 (2007): 1497–1508.

Fleurence, R., Selby, J.V., Odom-Walker, K., Hunt, G., Meltzer, D., Slutsky, J.R., et al. How the patient-centered outcomes research institute is engaging patients and others in shaping its research agenda. *Health Affairs* 32 (2013): 393–400.

Golub,T. Counterpoint: data first. *Nature* (2010) 464 (2010): 679. doi: 10.1038/464679a

Goodford, P.J. A computational procedure for determining energetically favorable binding sites on biologically important macromolecules. *Journal of Medicinal Chemistry* 28, no. 7 (1985): 849–857.

Gopisankar, M.G. YP2D6 pharmacogenomics. *Egyptian Journal of Medical Human Genetics* 18, no. 4 (2017): 309–313.

Healthcare Analytics Market to Hit $31 Billion by 2022. *Healthcare Finance News. 2018-08-13*. Retrieved February 5, 2019-02-05.

Hersh,W. A stimulus to define informatics and health information technology. *BMC Medical Informatics and Decision Making* 9(2009): 24.

Hudson, K.L., and Collins, F.S. Bringing the common rule into the 21st century. *The New England Journal of Medicine* 373 (2015): 2293–2296.

Jing, Y., Bian, Y., Hu, Z., Wang, L., and Sean Xie, X.-Q. Deep learning for drug design: an artificial intelligence paradigm for drug discovery in the big data era. *The AAPS Journal* 20, no. 3 (2018): 58.

Kalinin, S.V., Strelcov, E., Belianinov, A., Somnath, S., Vasudevan, R.K., Lingerfelt, E.J., Archibald, R.K., Chen, C., Proksch, R., Laanait, N., and Jesse, S. Big, deep, and smart data in scanning probe microscopy. *ACS Nano* 10, no. 10 (2016): 9068–9086. doi: 10.1021/acsnano.6b04212

Karczewski, K.J.,Daneshjou, R., and Altman, R.B. Pharmacogenomics, *PLoS Computational Biology* 8, no. 12 (2012): e1002817.

Keusch,G.T. What do-omics mean for the science and policy of the nutritional sciences? *The American Journal of Clinical Nutrition* 83 (2006): 520–522.

Kinnings, S.L., Liu, N., Buchmeier, N., Tonge, PJ., and Xie, L. Drug discovery using chemical systems biology: repositioning the safe medicine Comtan to treat multi-drug and extensively drug resistant tuberculosis. *PLOS ComputationalBiology* 5 (2009): e1000423.

Kola, I., and Landis, J. Can the pharmaceutical industry reduce attrition rates?*Nature Reviews Drug Discovery* 3 (2004): 711–715.

Kulynych,J., and Greely, H. Every patient a subject: When personalized medicine, genomic research, and privacy collide, 2014, http://www.slate.com/articles/technology/future_tense/2014/12/when_personalized_medicine_genomic_research_and_privacy_collide

Leelananda, S.P., and Lindert, S. Computational methods in drug discovery.*Beilstein Journal of Organic Chemistry* 12, no. 1 (2016): 2694–2718.

Lotfi Shahreza, M., Ghadiri, N., Rasoul Mousavi, S., Varshosaz, J., and Green, J.R. A review of network-based approaches to drug repositioning.*Briefings in Bioinformatics* 19, no. 5 (2018): 878–892.

Luce, B.R., Kramer, J.M., Goodman, S.N., Connor, J.T., Tunis, S., Whicher, D., et al. Rethinking randomized clinical trials for comparative effectiveness research: the need for transformational change. *Annals of Internal Medicine* 151 (2009): 206–209.

Luscombe, N.M., Greenbaum, D., and Gerstein, M. What is bioinformatics? A proposed definition and overview of the field. *Methods of Information in Medicine* 40, no. 4 (2001): 346–358.

Manyika, J., Chui, M., Brown, B., Bughin, J., Dobbs, R., Roxburgh, C., et al. *Big Data: The Next Frontier for Innovation, Competition, and Productivity [Internet]*. New York: McKinsey and Company; 2011 May. [cited 2014 May 28]. Available from: http://www.mckinsey.com/insights/business_technology/big_data_the_next_frontier_for_innovation

Margolis, R., Derr, L., Dunn, M., Huerta, M., Larkin, J., Sheehan, J., Guyer, M., and Green, E.D. The National Institutes of Health's Big Data to Knowledge (BD2 K) initiative: capitalizing on biomedical big data. *Journal of the American Medical Informatics Association* 21, no. 6 (2014): 957–958.

Min Jou, W., Haegeman, G., Ysebaert, M., and Fiers, W. Nucleotide sequence of the gene coding for the bacteriophage MS2 coat protein. *Nature* 237 (1972): 82–88.

Mnih, V., Kavukcuoglu, K., Silver, D.,Rusu, A.A., Veness, J., Bellemare, M.G., Graves, A., et al. Human-level control through deep reinforcement learning. *Nature* 518, no. 7540 (2015): 529–533.

Nagarajan, P. An over view of bioinformatics. *Trends in Biomaterials and Artificial OrgansTrends* 17, no. 2 (2004): 4–8.

Oldenburg, J., Watzka, M., Rost, S., and Müller, C.R. VKORC1: molecular target of coumarins. *Journal of Thrombosis and Haemostasis* 5, no. 1 (2007): 1–6.

Owen, R.P., Altman, R.B., and Klein, T.E. Pharm GKB and the International Warfarin Pharmacogenetics Consortium: the changing role for pharmacogenomics databases and single-drug pharmacogenetics. *Human Mutation* 29 (2008): 456–460.

Panteleev, J., Gao, H., and Jia,L. Recent applications of machine learning in medicinal chemistry. *Bioorganic & Medicinal Chemistry Letters* 28, no. 17 (2018): 2807–2815.

Qazi, S., and Raza, K. (2020). *Smart Biosensors for an EfficientPoint of Care (PoC) Health Management. Editors: Jyotismita Chaki Nilanjan Dey Debashis De, In Smart Biosensors in Medical Care* (pp.65–85). Elsevier. https://doi.org/10.1016/B978-0-12-820781-9.00004-8

Raza, K. Formal concept analysis for knowledge discovery from biological data. *International Journal of Data Mining and Bioinformatics, Inderscience* 18, no. 4 (2017): 281–300. https://doi.org/10.1504/IJDMB.2017.10009312

Raza, K., and Ahmad, S. (2019).Recent advancement in next-generation sequencing techniques and its computational analysis. *International Journal of Bioinformatics Research and Applications, Inderscience* 15, no. 3 (2019): 191–220. https://dx.doi.org/10.1504/IJBRA.2019.10022508

Roses, A.D. Pharmacogenetics and drug development: the path to safer and more effective drugs. *Nature Reviews Genetics* 5 (2004): 645–656.

Sellwood, M.A., Ahmed, M.,Segler, M.H.S., and Brown, N. Artificial intelligence in drug discovery. *Future Medicinal Chemistry* 10, no. 17 (2018): 2025–2028.

Senthilkumar, S.A., Rai, B.K., Meshram, A.A., Gunasekaran, A., and Chandrakumarmangalam, S. Big data in healthcare management: a review of literature. *American Journal of Theoretical and Applied Business* 4, no. 2 (2018): 557–569.

Service, R.F. The race for the $1000 genome. *Science* 311 (2006): 1544–1546.

Shih, I.M., and Wang, T.L. Apply innovative technologies to explore cancer genome. *Current Opinion in Oncology* 17 (2005): 33–38.

Simpao, A.F., Ahumada, L.M., Gálvez, J.A., and Rehman, M.A. A review of analytics and clinical informatics in health care. *Journal of Medical Systems* 38, no. 4 (2014): 45.

Soetan, K.O., and Abatan, M.O. Biotechnology a key tool to breakthrough in medical and veterinary research. *Biotechnology and Molecular Biology Reviews* 3, no. 4 (2008): 88–94.

Srinivasa, R., Das, S., Nageswara, R., and Kusuma, K., 2009. Bioinformatics is a key life science R & D activity. *International Journal of Bioinformatics Research* 1, no. 2 (2009): 81–84.

Su, C.T., Wang, P.C., Chen, Y.C., and Chen, L.F. Data mining techniques for assisting the diagnosis of pressure ulcer development in surgical patients. *Journal of Medical Systems* 36, no. 4 (2012): 2387–2399.

Svetnik, V., Liaw, A.,Tong, C.,Culberson, J.,Sheridan, R.P., and Feuston, B.P. Random forest: a classification and regression tool for compound classification and QSAR modeling. *Journal of Chemical Information and Computer Sciences*43, no. 6 (2003): 1947–1958.

Topol, E.J. High-performance medicine: the convergence of human and artificial intelligence. *Nature Medicine* 25, no. 1 (2019): 44–56.

Tsipouras, M.G., Exarchos, T.P., Fotiadis, D.I., Kotsia, A.P., Vakalis, K.V., Naka, K.K., and Michalis, L.K. Automated diagnosis of coronary artery disease based on data mining and fuzzy modeling. *IEEE Transactions on Information Technology in Biomedicine* 12, no. 4 (2008): 447–458.

Wang, W., Yang, S.,Zhang, X., and Li,J. Drug repositioning by integrating target information through a heterogeneous network model. *Bioinformatics* 30, no. 20 (2014): 2923–2930.

World Health Organization, 2013. *Global Tuberculosis Report 2013*. World Health Organization, Washington, DC.

Yu, M.K., Ma, J., Fisher, J., Kreisberg, J.F., Raphael, B.J., and Ideker, T. Visible machine learning for biomedicine. *Cell* 173 (2018): 1562–1535. doi: 10.1016/j.cell.2018.05.056.

Yun, M.R., Han, S.J., Yoo, W.G., Kwon, T., Lee, S., Lee, J.S., and Kim, D.W., 2016. Draft genome sequence of Mycobacterium tuberculosis KT-0204, isolated in South Korea. *Genome Announcements* 4, no. 1 (2016): e01519–15.

Zagursky, R.J., and Russell, D., 2001.Bioinformatics: use in bacterial vaccine discovery. *Biotechniques* 31, no. 3 (2001): 636–659.

Zhang, X.W. Biomarker validation: movement towards personalized medicine. *Expert Review of Molecular DiagnosticsExpert Rev Mol Diagn* 7 (2007):469–471.

2 Next-Generation Sequence Analysis for Clinical Applications

Agnik Haldar and Ajay Kumar Singh
Central University of South Bihar

CONTENTS

2.1 INTRODUCTION

Next-generation sequencing (NGS) can be documented as a high-throughput technique of DNA sequencing in parallel which is able to generate high amount of data in a relatively short amount of time. NGS has been derived from Sanger sequencing technologies and also aims to rectify itself from the shortcomings of Sanger sequencing. Since slowly cementing its place as the gold standard for sequencing studies, NGS has expanded its repertoire and branched into the realm of clinical applications. Over the years, there have been a few iterations for carrying out genome sequencing studies.

2.1.1 First-Generation Sequencing

The first-generation sequencing is comprised of Sanger sequencing and Maxam–Gilbert sequencing.

Developed in 1977 by Frederick Sanger, the first generation of sequencing technologies was built on the principle of chain termination method or the dideoxynucleotide (dNTP) method, where the first genomes which were sequenced were from Phage X174, which had a length of 5,375 bases (Sanger et al., 1977). The procedure consists of utilizing a single strand of the double-stranded DNA to be used as a template for the sequencing. dNTPs are nucleotides which have been clinically modified, and they play a major role in the sequencing by being tagged uniquely for each of the bases of the DNA. These are used in a bid to inhibit the elongation of the nucleotide due to the fact that dNTPs cannot form phosphodiester bonds during DNA synthesis. The fragments are generated in four separate columns of a gel slab according to the respective DNA bases in a process termed as gel electrophoresis. After the procedure, the obtained fragment samples are visualized and analyzed using an imaging system preferably X-ray or UV light (Kchouk et al., 2017).

Over the years, researchers did improve the existing method; for example, in 1995, Applied Biosystem introduced an automated Sanger sequencing machine known as the Applied Biosystems (ABI) Prism 370. It utilized technologies such as capillary electrophoresis, terminators which were fluorescently labeled, and automated laser signal detection which is an improved method of detecting nucleotide sequences (Garrido-Cardenas et al., 2017; Kchouk et al., 2017). For decades, Sanger sequencing was considered as the gold standard for genome sequencing and is still being used in some capacity today. The reason being its ease of availability and accuracy. It helped inspire other sequencing methods such as pyrophosphate sequencing method and ligation enzyme method which were used as 454 and SOLiD (supported oligonucleotide ligation and detection) techniques, respectively, and had at their core the basic principle of using dNTP to inhibit DNA synthesis. The future generations of the sequencing were also made possible due to Sanger sequencing. However, one flaw which plagued the technique and made it face the test of time was its throughput. Throughput is defined as a function based on the reaction time of the sequencing, the amount of sequencing reactions possible to be performed in parallel, and the amount of sequence length being read by each reaction. Sanger sequencing provided a low throughput due to the electrophoretic separation, taking a lot of time and providing considerably less output. To add to that, the cost of Sanger sequencing is also quite high compared to other alternatives present today.

Maxam–Gilbert sequencing, also known as the chemical degradation method developed during 1976–1977 by Allan Maxam and Walter Gilbert, is another sequencing method considered among the first generation of genome sequencing (Maxam & Gilbert, 1977). The main principle adheres to modification of the DNA chemically and subsequently cleaving at specific bases. The chemical modifications aid in creating breaks at any one or two of the nucleotide bases. Depending on the combination of the chemicals used, different fragments are obtained. Even though the procedure and principle were quite straight forward, it lost its popularity due to the cost being too high, the procedure being time consuming, and the fact that it required the usage of toxic radioactive chemicals.

2.1.2 Second-Generation Sequencing

The beginning of 2005 and the following years marked the rise of the second generation of genome sequencing which was aimed at improving the throughput problem of the Sanger sequencing and also would mark the beginning of NGS. The directives governing the second generation of genome sequencing were a high-throughput result and the ability of obtaining sequencing output without the need of gel electrophoresis.

Some of the famous sequencing technologies developed during this period include Roche/454, Illumina/Solexa, and the ABI/SOLiD sequencing, with most of them based on the main principle of Sanger sequencing.

Rocher/454 sequencing was developed in 2005 and it used pyrosequencing, which is the detection of pyrophosphate after it is released due to the addition of each nucleotide to a new synthetic DNA strand. The principle it follows has been dubbed as the "sequencing by synthesis" (Harrington et al., 2013; Ronaghi, 2001). The light emitted after each nucleotide is incorporated is recorded, and the sequence of the DNA fragment is obtained. Even though 454 sequencing was able to sequence long reads, it was prone to errors resulting due to insertions and deletions. Also the signals that had an overtly high or low intensities resulted in errors while generating the sequences (Kchouk et al., 2017).

Illumina/Solexa sequencing follows the sequencing by synthesis approach and is currently one of the most used platforms for NGS. The first step in the sequencing is that the DNA is broken up in fragments and tags are attached to each end of the sequence. Next the primer, indices, and terminators are added to the sequence, and using polymerase chain reaction (PCR) the sequences are "amplified" or copied multiple times. In the final step, the Illumina determines each nucleotide in the sequence and finally by sequencing by synthesis the reversible terminators are used and the hybridization of the primer takes place. Implementation of the polymerases takes place and the primers are as a result extended. Each nucleotide is fluorescently labeled to differentiate among the four nucleotides. The signals the fluorescent light generates is recorded and translated into nucleotide sequence. The limitations of the process include the short length reads which can deter its usage in some specific cases.

SOLiD, acquired by ABI in 2007, works on the principle of ligation enzyme method. The core principle of the system is based on the Sanger sequencing principle of chain termination method. The process comprises several rounds of sequencing, starting from adapters being attached to the DNA fragments which are fixed on beads (Kchouk et al., 2017; Liu et al., 2012). These beads are then transferred to a glass surface. The higher the quantity of the beads, higher the throughput. The beads are then introduced to a library of 8-mer probes which contain separate fluorescent dyes and are eventually ligated to the DNA fragments. The intensity of the color emitted is recorded for analytical purposes (Table 2.1). The technique has been found to be extremely accurate and inexpensive; however, the drawback is that the read lengths are extremely small thus making their applications unfavorable in most conditions (Liu et al., 2012).

TABLE 2.1
Generation-wise DNA Sequencing Technologies

Generation	Name	Description
First generation	Sanger sequencing	Developed in 1977, based on the principle of chain termination method or the dideoxynucleotide method
	Maxam–Gilbert sequencing	Developed during 1976–1977, principle adheres to modification of the DNA chemically and subsequently cleaving at specific bases
Second generation	Roche/454 sequencing	Appeared in market in 2005, used pyrosequencing which is the detection of pyrophosphate after it is released due to the addition of each nucleotides to a new synthetic DNA strand
	Ion torrent sequencing	Developed in 2010, similar to 454 pyrosequencing; however, it is based on the detection of the hydrogen ion released during the sequencing process
	Illumina/Solexa sequencing	Sequencing by synthesis approach and is currently one of the most used platforms for next-generation sequencing (NGS)
	Applied Biosystems (ABI)/ supported oligonucleotide ligation and detection (SOLiD) sequencing	Acquired by ABI in 2007, works on the principle of ligation enzyme method
Third generation	Pacific biosciences SMRT sequencing	The first genomic sequencer using SMRT approach but high error rates
	Oxford Nanopore sequencing	Developed in 2014 to determine the order of the nucleotides. Related to MinION which promises longer reads and better structural resolution. However, error rates are too high

2.1.3 Next-Generation Sequencing

In the field of genomic research, the advent of next-generation sequencing (NGS) is marked as a revolutionary achievement paving the way for unprecedented technology for genome sequence analysis. They are also termed as massively parallel or deep sequencing as these are considered umbrella terms for defining DNA/RNA sequencing technologies. NGS in comparison to Sanger sequencing was able to sequence the entire genome in a single day, whereas Sanger sequencing took almost a decade. Thus, the time efficiency aspect is unparalleled in terms of any present sequencing methods.

The main steps involved in a typical NGS experiment include DNA/RNA preparation, library preparation by indexing the DNA or RNA segments, sequence analysis, bioinformatics approaches which deal with the raw data obtained, data and statistical analysis, and finally mutation annotation analysis which leads to the possible biological or clinical significance. NGS sequencing is known for its unprecedented speed and throughput; however, it is not free from some limitations. For example, the high throughput of NGS sequencing comes at the price of having a shorter read length. The most in demand platforms like Illumina or SOLiD can only manage an average read length of 30–400 bp as compared to Sanger's sequencing which can manage a read length of up to 1 kb. There are a few third-generation sequencing methods in development which have shown promise in terms of speed as well as read length; however, they are not readily available and are not at the completion stage (Liu et al., 2012).

Shorter read lengths can cause limitations to the usage of NGS in certain experiments, for example, it becomes increasingly difficult to do de novo genome assembly using shorter reads. Therefore, to counter this problem somewhat, a process called resequencing is used which basically uses a template genome to map those shorter reads. That process too can lead to problems if the short reads do not align properly with the template genome. Things also tend to become difficult when factors such as structural variants and region diversities are involved. These problems for now are tackled using longer read lengths and paired end sequencing which allows sequencing at both ends of the fragment in a bid to increase its accuracy (Hartman et al., 2019; Siu et al., 2015).

The reads obtained after sequencing are generally in their raw format, as such the massive data it generates is not of much use in that state. Therefore, the data obtained is used to perform statistical and data analysis to better understand the nature and relations of the genomes accordingly. NGS has the capacity to generate data sets reaching up to gigabases (billions). To better understand such huge amount of data a number of statistical procedures have been used to a greater effect.

There are other methods such as the Pacific Biosciences SMRT (Single-molecule real-time) sequencing or the Oxford Nanopore sequencing, which have had soft releases but are still in need of further development due to the tools not being as stable as its predecessors. Using nanopore sequencing technology, Oxford Nanopore sequencing has developed MinION, a portable USB-like device which can be connected to a computer or laptop, and it is capable of understanding and translating the DNA sequence data of an individual into the respective nucleotides. This data can then be made accessible via streaming into cloud servers and in turn to researchers and scientists alike. Dubbed as the advent of third-generation sequencing, these methods hold a high potentiality to revolutionize the healthcare sectors due to its functionality and ability to detect mutations of deadly viruses and thus mitigate any outbreaks (Raza & Qazi, 2019). However, the error rates of the sequencers are high when compared to its predecessors. Nonetheless, the tools are still in their developmental phases and have shown sufficient promise to be heralded as the new third generation of sequencing methods.

2.2 CLINICAL APPLICATIONS

The application of NGS in a clinical setting is still in its infancy, as a result of which the topic is not yet fully explored in the area. However, it does not change the potential that NGS has in the clinical setting, and to back it up the work being done presently has shown tremendous results. NGS becomes extremely useful in cases where the need arises for a huge number of sample tissues for molecular assays, where it helps analyze the required targets in a single run, thus saving the need for multiple tissues for multiple assays. The experimental cost can be reduced drastically as well as the time taken for the entire experiment (Meldrum et al., 2011). The procedure is especially seen to be beneficial for a number of diseases, most specifically cancer. The clinical approach and possibilities for each type of diseases are discussed below.

2.2.1 Cancer

Being a genetic disorder which is generally a result of hereditary or somatic mutations, NGS can help in revolutionizing DNA sequencing technologies to better detect, treat, and manage the disease efficiently. Currently, The Cancer Genome Atlas project, which began in 2005, utilizes NGS to catalogue genetic mutations which are responsible for a variety of cancers. This in turn will help the research community better understand the mechanisms behind disease pathogenicity, thus paving a path for personalized medicines based on molecular pathology (Seifi et al., 2017). Potentially, the advantages of implementing NGS methods for managing genetic diseases like cancer are quite a few. The avenues these methods open in the treatment of cancer ranges from personalized treatment to detection and availability of useful information which can lead to the development of new drug targets.

Over the years, one of the norms for cancer diagnostics was the analysis of cancer tissues by liquid biopsy (Diaz Jr & Bardelli, 2014; Savelieff, 2019; Siu et al., 2015). When tumors are metastasized in the body, they tend to leave a trail of circulating cancer cells and circulating tumor RNA and DNA in the circulation system. Detection of these components were seen as the normal practice. However, with the advent of NGS, it built upon the drawbacks of the system having less sensitivity when detecting these components or being too limited in terms of detection, as seen in cases of digital droplet PCR where it only works for specific genes. The NGS system having rectified these shortcomings is able to detect all types of nucleic acid from known or unknown genes. The implementation of NGS in a clinical setting has also paved the way for a less invasive and faster method of diagnosis compared to the earlier techniques. In diseases like cancer where it is difficult to pin point the mutations seen in the carcinogenic genes and differentiate them from normal genes, NGS aims to solve the problem by sequencing the whole genome and detects the majority of the nucleic acids, thus proving its worth in biomarker discovery for cancer. There are also several reports which state the usage of NGS technology to detect target-specific somatic tumor mutations which can be particularly useful and advantageous to understand the extent and strategize according to the need of the patient (Giardina et al., 2018; Kim et al., 2019). The cancer genomes can be further explored by performing molecular profiling on them, which can lead to a better understanding of the different cancer types as well as perform molecular diagnosis. It was also reported that NGS methods performed at a much higher sensitivity while detecting somatic mutations when compared to Sanger sequencing (Arsenic et al., 2015). Other uses of NGS in clinical setting would be the identification of ctDNA which can help keep a track on disease progression and the identification of exosomal RNAs which can lead to the identification of noncoding RNAs, mRNAs, etc. as they are contained within the exosome and secreted by both normal and cancer cells. Thus any dysregulation would inform us of the mutated component (Siu et al., 2015).

2.2.2 Epigenetic Diseases

Epigenetics is defined as any heritable change witnessed in the gene expression however the DNA sequence is unaltered. This results in a drastic alteration in the nature of cells tasked with reading the genes. Given the current scenario, technological advancements in the field of genetics have advanced considerably and bioinformatics approaches are on the rise. Added to that, a plethora of diseases with epigenetic implications are beginning to surface as a result. Such diseases include cancer, reproductive diseases, autoimmune disorders, etc. The list of diseases linked with epigenetic regulation are increasing as research progresses, which as a result helps us understand the genes affecting the diseases and whether those genes can be manipulated to control the production or barring the production of certain proteins or enzymes in specific cells, which might end up proving beneficial in the course of tackling the disease. A famous example would be the methylation process which is an epigenetic function of the gene, inhibits the GCboxes (guanine-cytosine) by altering its transcriptional activation (Nomura et al., 2007). There are three main epigenetic processes which play a part in the cause of the disease: chromatin remodeling, histone modification, and DNA methylation. Understanding these modifications are crucial as they can act as epigenetic markers for gene expression and activity. Often epigenetics can be related with phenotype and genotypes as they can elucidate the mechanism of the cells having the same DNA can generate different cell types. It is where the potentiality of NGS technologies comes to the forefront. Epigenome can be explained as the chemicals and proteins which interact with the DNA to affect the genes. Mapping these epigenomic properties can lead to discovery or identification of biomarkers which can be helpful for various therapeutic treatment. The Human Epigenome Project, a multinational project, aims to do just that—which is "identify, catalog, and interpret genome-wide DNA methylation patterns of all human genes in all major tissues" (Eckhardt et al., 2004). In a clinical setting, the usage of methylated DNA immunoprecipitation and bisulfite methods along with ChIP-Seq (Chromatin immunoprecipitation sequencing) technology can help usher the information gained from DNA methylation and the location of transcription factors, both of which has the potential to lead us to new biomarkers. ChIP-Seq on the NGS platform dubbed ChIP-chip, which is a combination pipeline of ChIP-Seq and microarray, and it is being used to study protein–DNA interactions (Seifi et al., 2017).

2.2.3 Mendelian Diseases

Mendelian diseases are genetic disorders born of mutation of a single gene and they include diseases such as cystic fibrosis, albinism, sickle cell anemia, and thalassemia. Mendelian or monogenic disorders can be divided into five types: autosomal dominant, autosomal recessive, sex-linked dominant, sex-linked recessive, and mitochondrial. Sanger sequencing is considered as the method of choice in clinical environment when it comes to analyzing mendelian diseases by confirming a diagnostic element related to it and providing appropriate genetic counseling. However, NGS is slowly being introduced in clinical studies due to its cost effectiveness and it takes much less time as compared to Sanger sequencing. It is being used in cases of suspected mendelian diseases to identify rare variants in patients. Miller's syndrome was one of the first rare mendelian diseases for which rare variants were identified as a result of whole exome sequencing (WES) (Ng et al., 2010; Seifi et al., 2017). There were other uses of WES, which was employed for identifying somatic and rare mutations in tumors and later on helped in clinical diagnoses as well. The usefulness of each of the methods varies according to the types of disorders we encounter. The clinical therapy and diagnostics do not necessarily change as a result of the findings; however, the cost and time can be considerably managed accordingly along with a possibility of unnecessary complications being avoided completely. Gene-specific therapy can be made possible as a result of such procedures which can help identify causative variants across patients. This helps in a better management strategy as well as reducing the chance of any mistreatment (Jamuar & Tan, 2015).

Further identifying the molecular etiology would benefit the clinicians in providing an accurate estimate of factors like chance of recurrence, risk of the disease being hereditary, and strategizing accordingly to counter it in its presymptomatic stages (Biesecker & Green, 2014). An example would be the identification of the mutated gene in long QT syndrome (Theilade et al., 2013). It can help identify potential risks to family members and develop appropriate strategy to avoid such mishaps. NGS also has the potential for identifying gene–gene interactions which might help us develop novel techniques for targeted molecular therapy in future. There has been reports of such procedures in various cases of cancer (Malone et al., 2020; Saito et al., 2018).

The current objective of the whole genome association surveys is to concentrate on single-nucleotide polymorphism and their genetic effects in diseases with the help of microarray technologies (*Whole Genome Association Studies*, 2011). Factors such as rare and copy number variants might have crucial implications on disease phenotypes.

2.2.4 Infectious Diseases

The application of NGS, for the purpose of viral drug resistance testing, has been explored in recent years quite successfully. Particularly, given the current scenario of the world due to the COVID-19 outbreak, NGS methods in clinical virology can benefit exponentially in terms of speed and accuracy. Drug resistance forms an important pillar in mitigating the effects of a plethora of clinically diagnosed viral infections. At first, Sanger sequencing was the method of choice for detecting genetic aberrations. However with the advent of latest NGS technologies, they replaced the Sanger sequencing as the preferred method for detecting mutations pertaining to drug resistance. A major reason for it was that Sanger sequencing was unable to detect minor variants when they were present in a small quantity. Whereas NGS methods were able to detect such variants even when they were present at around 1% in the viral population.

One of the most common viruses which is used extensively for such genotypic resistance testing is HIV-1. Several research has pointed out the fact that HIV-1-afflicted patients, when epidemiologically tested, showed that mutations which facilitate forming resistance of the virus to antiretroviral therapy are present in the genetic makeup of the patient. Thus, a genotypic testing is quite necessary for patients as it can help confirm the drug-resistant mutations present in the patient as well as survey the population for such mutations (Vrancken et al., 2010). DEEPGENHIV (Gibson et al., 2014) is a genotypic assay developed to measure the susceptibility of the HIV-1 with all the drugs which can affect the viral enzymes present in the virus. It also serves the purpose of speculating the co-receptor tropism and has the provision to analyze 96 samples in a single run. Other clinically approved assays for detecting HIVDR include MiSeq-HyDRA (Taylor et al., 2019), which takes the properties of IlluminaMiSeq and merges it together with HyDRA web server, PASeq, a cloud-based HIV genotype testing web service aimed at a fully automated analysis of HIV genotype data using NGS-based methods. The abovementioned tools are used in a bid to identify and quantify amino-acid variants (AAV) of HIV. AAVs can be explained as any amino acid changes which occur in the HIV-1 reference sequence. The abovementioned tools were validated to perform quite efficiently in detecting such variants (Lee et al., 2020).

2.3 THE SCOPE OF NGS TECHNOLOGY IN CLINICAL ENVIRONMENT

Using a specific technology in clinical setting needs to be adequately decided and regulated in a bid to stop its misuse. To facilitate that, several official guidelines have been implemented which have been overseen and decided by several associations. These guidelines serve as a systematic procedure for experiments to proceed without any hindrance.

The application of NGS for clinical laboratories in the current scenario can either be a targeted analysis or a whole genome/exome/transcriptome sequencing analysis. In case of targeted genes, it deals with an approximate of a few thousand base cells, whereas whole genome sequencing (WGS) can manage the analysis of around 3.3 billion bases, and WES tackles around 40–50 million bases. The difference between targeted gene analysis and WES/WGS lies in the fact that the latter can identify mutations in the genome along with the unknown variants; in contrast targeted gene analysis works on detecting specific genes. Targeted gene analysis also provides an exon coverage which is greater than WGS and WES applications. For clinical purposes, targeted gene analysis is the preferred option. WGS and WES are mainly used for research studies (Qin, 2019). However, there are studies being conducted where the possibility of WES being regarded as a clinical analysis process is being explored, which can impact medical management and treatment (Niguidula et al., 2018). Interestingly, another candidate for NGS in clinical settings would be whole transcriptome sequencing. Though not yet quite common as compared to its other counterparts, the use of microarrays and RNA-sequencing (RNA-seq) are slowly being incorporated in clinical environments for identification of novel disease markers and diagnosis at a molecular level. When dealing with diseases like cancer, a lot of the information can be found in its genetic study. Microarrays and RNA-seq experiments allow for the same. The genes which are found to not perform their usual task and they appear as a result of some mutations are dubbed as dysregulated genes. The identification of these genes can often help us narrow down and explore important and novel biomarkers. Transcriptomic sequencing paired with transcriptomic analysis can open uncharted territories which have been made possible owing to the advancement in technology with the advent of robust and more powerful computational resources (Casamassimi et al., 2017). The data generated by microarray or RNA-seq experiments are compiled as raw data and then analyzed following distinct RNA-seq or microarray data analysis workflows coupled with machine learning algorithms, which point us in the direction of our desired objective(Casamassimi et al., 2017; Jabeen et al., 2018). There have been several projects highlighting its use in cancer research studies ranging from identification of protein coding genes to detection and analysis of non-coding RNAs, all of which has seen much success. The effect of noncoding RNAs has been seen at an epigenetic level along with novel genetic mutations which results in oncogenesis and cancer progression. NGS-based research has made it possible and has provided an answer for diagnosis at a molecular level for a life-threatening disease like cancer and hopefully concentrate on providing precision treatment to patients (Figure 2.1).

2.4 PROCEEDINGS OF NGS METHODS IN CLINICAL APPLICATION

Before embarking on a project in either clinical or research environment, a proper laid out plan needs to be decided upon to minimize the risk of errors and other complications. Similarly in a clinical setting, the usage of NGS needs a workflow to follow (Figure 2.2).

The first step comprises, the genetic mutation to be studied, identified and decided upon. This is usually the base on which the whole experiment will vary and stand upon. In case of any disease, the protocol suggests to check the guidelines for the selected disorder or disease and sort out the gene mutations already present. For instance, if we take the example of a disease which is quite actively researched like cancer, new findings are being updated almost every day. Thus, the need for an extensive literature survey gains precedence in a bid to identify the targets yet to be discovered (Qin, 2019).

The second step in the workflow would be the formulation of a protocol based specifically on the chosen disease. There exists a number of different methods for the same disease and having the same end goal; however, depending on the patient's history and prognosis, the procedure needs to be molded accordingly. Some of the options like amplicon assay and hybrid capture assays exist to provide options for our needs. For example, amplicon assay can be completed within a short period of time but hybrid capture assay has a higher throughput than amplicon (Table 2.2). Basically, every procedure will have its pros and cons, but the correct methods to be selected will depend on the situation at hand (DeWitt, 2019; Qin, 2019).

After deciding on a method, the next step would be to create a standard operation procedure (SOP). An SOP is a step-by-step list of instructions which needs to be followed. Along with writing an SOP, the experimental steps also need to be checked to make sure the steps are foolproof.

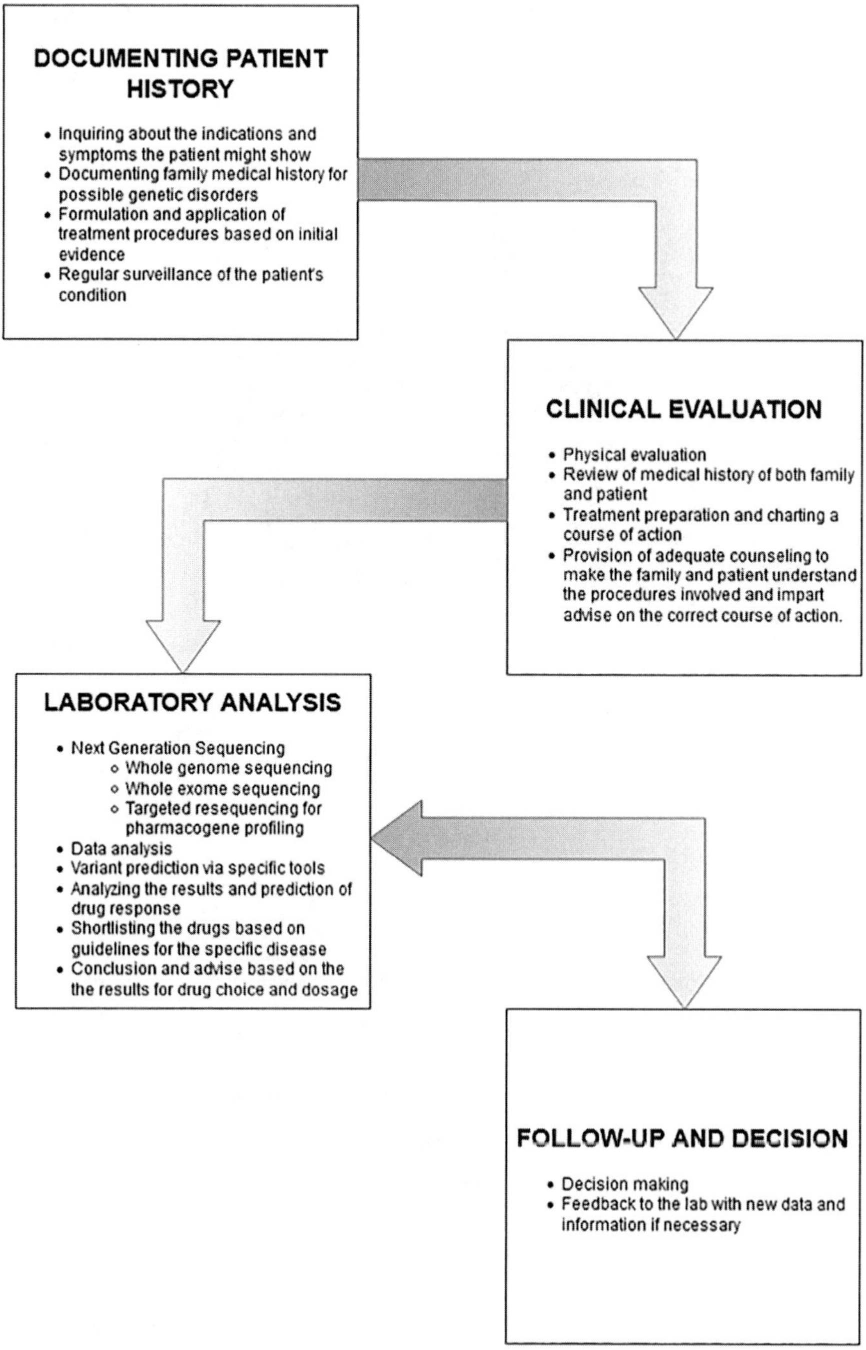

DOCUMENTING PATIENT HISTORY

- Inquiring about the indications and symptoms the patient might show
- Documenting family medical history for possible genetic disorders
- Formulation and application of treatment procedures based on initial evidence
- Regular surveillance of the patient's condition

CLINICAL EVALUATION

- Physical evaluation
- Review of medical history of both family and patient
- Treatment preparation and charting a course of action
- Provision of adequate counseling to make the family and patient understand the procedures involved and impart advise on the correct course of action.

LABORATORY ANALYSIS

- Next Generation Sequencing
    - Whole genome sequencing
    - Whole exome sequencing
    - Targeted resequencing for pharmacogene profiling
- Data analysis
- Variant prediction via specific tools
- Analyzing the results and prediction of drug response
- Shortlisting the drugs based on guidelines for the specific disease
- Conclusion and advise based on the the results for drug choice and dosage

FOLLOW-UP AND DECISION

- Decision making
- Feedback to the lab with new data and information if necessary

FIGURE 2.1 Next-generation sequencing methods in clinical setting.

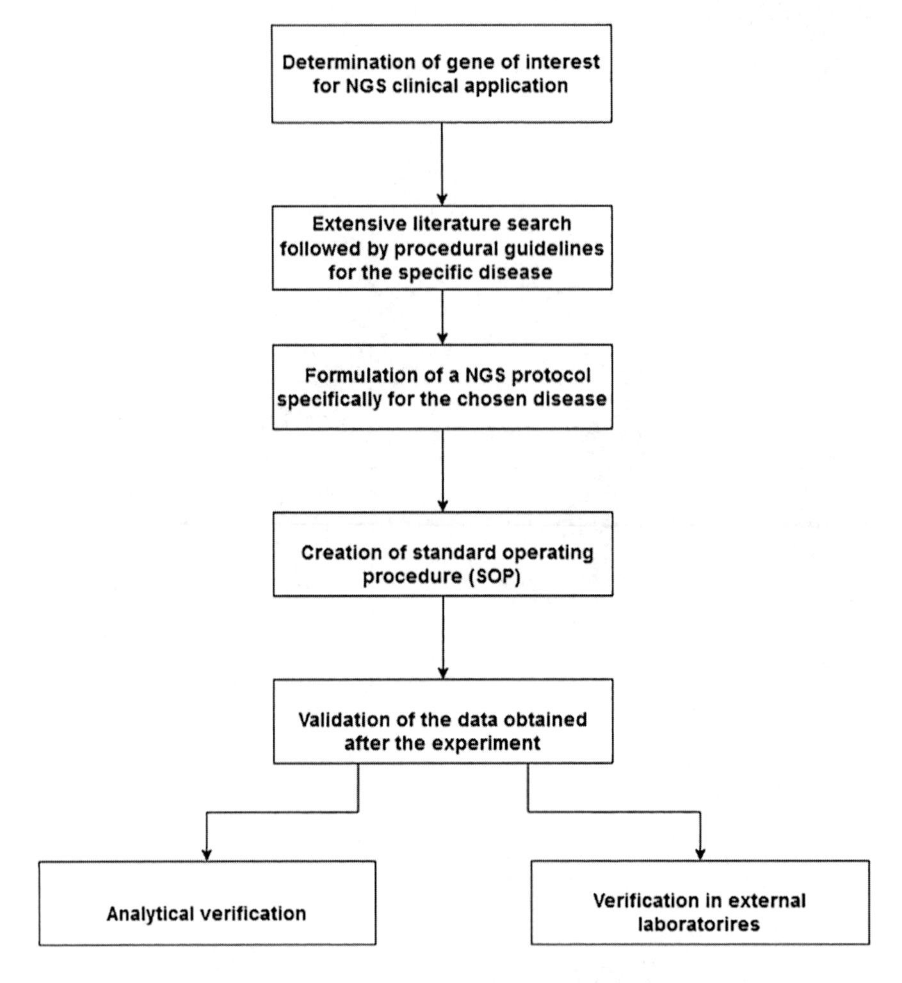

FIGURE 2.2 Workflow of next-generation sequencing methods in clinical setting.

TABLE 2.2
Comparison Between Amplicon Assay and Hybrid Capture Assay

Features	Amplicon Assay	Hybrid Capture Assay
Time taken for completion	Comparatively less amount of time	Comparatively more amount of time
Throughput	Comparatively low	Comparatively high
Cost per sample	Low cost	Can vary depending on the sample
Total number of steps	Comparatively less number of steps	Comparatively more number of steps
Number of targets per panel	Unlimited	9,000–10,000 amplicons

2.5 VALIDATION OF THE RESULTS OBTAINED

The last step requires validation of the data obtained after performing the experiments. The assays need to be verified in external laboratories to assure their accuracy, reproducibility, sensitivity, etc. Analytical validation is also a possibility which can help verify the range of the assay. According to regulatory requirements, laboratories which are certified by Clinical Laboratory Improvement Amendments are allowed to check the assays. The first step in the validation process requires the statement of intended use which needs to be explicitly declared. Often the validation protocol statement is required to be approved by the laboratory director before starting the validation process (Jennings et al., 2017).

The validation process starts with the samples which have been specified in the protocoland needs to be checked by two different people. To check for the reproducibility, it also needs to be tested at different runs.

An important parameter in the validation process would be the limit of detection (LOD). LOD can be explained as the absolute lowest value possible for a substance that can differentiate the substance from being completely absent or barely present which is denoted by a confidence level. Further defining LOD as lower limit of detection (LLOD) can elucidate the lowest variant allele frequency (VAF). Therefore, if the value of LLOD is less, the sensitivity of the samples would be high (Jennings et al., 2017; Qin, 2019). In case of NGS assays, the sequencing depth is another important factor which helps us understand the quality of the reads obtained. Higher the sequencing depth, purer the reads. It can be further differentiated into two terms, average and minimal depths, with minimal depth being defined as the minimum sequencing depth required for accurate results. The allelic frequency is cutoff, another parameter which helps in the identification of real variants from false variants. These are all included in the quality control matrix which needs to be monitored efficiently (Jennings et al., 2017; Kadri et al., 2017; Mallaret et al., 2016; Qin, 2019).

Finally, the data obtained after testing and validation needs to be interpreted. NGS usually identifies a plethora of variants; however, their significance clinically cannot be determined fully due to the variants being largely unknown or rarely reported. This makes the interpretation of the results tricky. To counter this, the most logical steps to take would be to follow the guidelines of the said disease, being updated frequently about the latest discoveries in the field as they may guide the interpretation toward the desired clinical significance, and finally scourge the literature databases available for literature review which might help connect the dots with biological and clinical significance.

2.6 LIMITATIONS AND FUTURE PROSPECTS

The NGS has revolutionized the research community with unprecedented technology pertaining to DNA sequencing. However, the procedure is not devoid of some drawbacks. One of the most glaring and obvious drawback of the procedure would be the difficulty in interpreting the data obtained and making it clinically significant. One of the reasons would be the lack of variant interpretation due to the presence of thousands of single-nucleotide variants which do not show any pathogenic relationships in addition to the fact that the variants which are discovered are rare and unknown (dubbed as variant of unknown significance) most of the time (Di Resta et al., 2018). This makes it hard to follow-up on relations without any concrete proof to pathogenic or physiological pathways. For a molecule or gene to reach up to the clinical stage, it needs to be validated experimentally and cross-checked for any discrepancies. However, the genetic information obtained was used as a confirmation for a particular process, and it hardly warrants a decisive role required for diagnosis. Therefore, the need to convert this genetic information into clinically significant data is of paramount importance.

The large amount of data NGS analysis generates is also a point which needs to be addressed. The need for high computational power and storage capacity is of utmost importance when it comes to NGS data analysis. Bioinformatics plays an important part in connecting and interpreting the raw data into statistically significant data which can be of use in clinical trials and laboratories. NGS data analysis has the capacity to narrow down prospective genes for further analysis. Instead of analyzing millions of genes, a bioinformatics approach reduces that number significantly bringing it down to a few hundreds (Raza & Ahmad, 2019; Shen et al., 2015). As of the year 2020, NGS methods have been successfully introduced in a clinical setting and is in fact efficient in identification of various biomarkers which can help in either prognosis or diagnosis of a disease, evaluate the extent of the reaction to a therapy, predict the chances of recurrence in cancer, and check the presence of a particular disorder or disease in the organism (Baldo, 2020).

The current objective to eliminate these drawbacks would be to concentrate on developing a primary diagnostic pipeline which would covert the phenotypic data to clinically and pathogenically significant. The NGS, owing to its fast, cost-effective, and reliable sequencing has the potential of becoming one of the spearheads of clinical study. The future of NGS in a clinical setting looks bright as it is evident from the success it has received in the field. It has been predicted to be an integral component in personalized treatment and deciding the correct course of action for therapies pertaining to various diseases. For the case of cancer treatment, it has been speculated that NGS will allow clinicians to predict their course of action for therapy based on the results of the sequencing data obtained for specific types of tumors. Pharmacogenomics is another avenue that the NGS methods in a clinical setting has opened for clinicians. It can help understand the workings behind the gene's response to a particular drug in addition to the fact that variant identification based on genomic sequencing of the normal and affected tissues will help us identify dysregulated genes and pave the way to understand the effects of those said variants on the toxicity or efficacy of certain drugs or molecules (Baldo, 2020; Kohlmann et al., 2012). The application of NGS in a clinical environment is still being fully adopted and so far has seen impressive results; however, the abovementioned limitations have been holding it back from bursting into the research scene. The advent of third and fourth generation of sequencing is already met with positive reactions among the research and clinical community, with the technology being implemented slowly in clinical studies.

2.7 CONCLUSION

The application of NGS in a clinical setting still remains a challenge today. NGS has been completely capable of sequencing whole genome and obtaining majority of the variants. The problem arises with the variants which are unknown as they cannot be annotated or related with a function. To counter this problem, analysis of more phenotypes with genotypes and mutations needs to be done. The more data gets annotated, the less number of unknown variants there would be. Another factor which might affect the application would be adequate computational architecture. The sheer amount of data which NGS generates without proper storage capacity and without enough computational power would be impossible to continue. The advent of NGS ushered a new era of DNA sequencing. Now, the sequencing generation has entered its fourth stage with single-cell sequencing. It has been seen to give high-throughput analysis and especially useful for analysis of variability for cancer cells. This new technique is called spatial transcriptomics. NGS for clinical diagnostic studies is still in its infancy; as a result the standard quality control of the process is still under major development. The accuracy as well as integrity of the data obtained need to be validated via proper quality control protocol. The application of NGS in clinical setting is well on its way progressing at a steady rate; however, this is not without considerable and serious limitations which is stopping the procedure from reaching its full potential. Overcoming those limitations would make NGS methods a common name in the world of clinical diagnostics.

REFERENCES

Arsenic, R., Treue, D., Lehmann, A., Hummel, M., Dietel, M., Denkert, C., & Budczies, J. (2015). Comparison of targeted next-generation sequencing and Sanger sequencing for the detection of PIK3CA mutations in breast cancer. *BMC Clinical Pathology, 15*, 20–20. PubMed. https://doi.org/10.1186/s12907-015-0020-6.

Baldo, L. (2020). *Impact of Next-Generation Sequencing Tests on Clinical Pathways for Cancer Care*.https://www.ajmc.com/journals/evidence-based-oncology/2020/february-2020/impact-of-nextgeneration-sequencing-tests-on-clinical-pathways-for-cancer-care.

Biesecker, L. G., & Green, R. C. (2014). Diagnostic clinical genome and exome sequencing. *New England Journal of Medicine, 370*(25), 2418–2425.

Casamassimi, A., Federico, A., Rienzo, M., Esposito, S., & Ciccodicola, A. (2017). Transcriptome profiling in human diseases: new advances and perspectives. *International Journal of Molecular Sciences, 18*(8), 1652. PubMed. https://doi.org/10.3390/ijms18081652.

DeWitt, J. (2019). Targeted sequencing: hybridization capture or amplicon sequencing? *Features and Applications of Targeted next Generation Sequencing*. https://eu.idtdna.com/pages/education/decoded/article/targeted-sequencing-hybridization-capture-or-amplicon-sequencing.

Di Resta, C., Galbiati, S., Carrera, P., & Ferrari, M. (2018). Next-generation sequencing approach for the diagnosis of human diseases: open challenges and new opportunities. *EJIFCC, 29*(1), 4–14. PubMed.

Diaz Jr, L.A., & Bardelli, A. (2014). Liquid biopsies: genotyping circulating tumor DNA. *Journal of Clinical Oncology, 32*(6), 579.

Eckhardt, F., Beck, S., Gut, I. G., & Berlin, K. (2004). Future potential of the human epigenome project. *Expert Review of Molecular Diagnostics, 4*(5), 609–618. https://doi.org/10.1586/14737159.4.5.609.

Garrido-Cardenas, J. A., Garcia-Maroto, F., Alvarez-Bermejo, J. A., & Manzano-Agugliaro, F. (2017).DNA sequencing sensors: an overview. *Sensors (Basel, Switzerland)*, *17*(3), 588. PubMed. https://doi.org/10.3390/s17030588.

Giardina, T., Robinson, C., Grieu-Iacopetta, F., Millward, M., Iacopetta, B., Spagnolo, D., & Amanuel, B. (2018).Implementation of next generation sequencing technology for somatic mutation detection in routine laboratory practice. *Pathology*, *50*(4), 389–401.

Gibson, R. M., Meyer, A. M., Winner, D., Archer, J., Feyertag, F., Ruiz-Mateos, E., Leal, M., Robertson, D. L., Schmotzer, C. L., & Quiñones-Mateu, M. E. (2014).Sensitive deep-sequencing-based HIV-1 genotyping assay to simultaneously determine susceptibility to protease, reverse transcriptase, integrase, and maturation inhibitors, as well as HIV-1 coreceptor tropism. *Antimicrobial Agents and Chemotherapy*, *58*(4), 2167–2185. PubMed. https://doi.org/10.1128/AAC.02710-13.

Harrington, C. T., Lin, E. I., Olson, M. T., & Eshleman, J. R. (2013).Fundamentals of pyrosequencing. *Archives of Pathology and Laboratory Medicine*, *137*(9), 1296–1303.

Hartman, P., Beckman, K., Silverstein, K., Yohe, S., Schomaker,M., Henzler, C., Onsongo, G., Lam, H. C., Munro, S., Daniel, J., Billstein, B., Deshpande, A., Hauge, A., Mroz, P., Lee, W., Holle, J., Wiens, K., Karnuth, K., Kemmer, T., … Thyagarajan, B. (2019). Next generation sequencing for clinical diagnostics: five year experience of an academic laboratory. *Molecular Genetics and Metabolism Reports*, *19*, 100464.https://doi.org/10.1016/j.ymgmr.2019.100464.

Jabeen, A., Ahmad, N., & Raza, K. (2018).Machine Learning-Based State-of-the-Art Methods for the Classification of RNA-Seq Data. In N. Dey, A. S. Ashour, & S. Borra (Eds.), *Classification in BioApps: Automation of Decision Making* (pp. 133–172). New York, NY: Springer International Publishing. https://doi.org/10.1007/978-3-319-65981-7_6.

Jamuar, S. S., & Tan, E.-C. (2015). Clinical application of next-generation sequencing for Mendelian diseases. *Human Genomics*, *9*(1), 10. https://doi.org/10.1186/s40246-015-0031-5.

Jennings, L. J., Arcila, M. E., Corless, C., Kamel-Reid, S., Lubin, I. M., Pfeifer, J., Temple-Smolkin, R. L., Voelkerding, K. V., & Nikiforova, M. N. (2017).Guidelines for validation of next-generation sequencing–based oncology panels: a joint consensus recommendation of the association for molecular pathology and college of American pathologists. *The Journal of Molecular Diagnostics*, *19*(3), 341–365. https://doi.org/10.1016/j.jmoldx.2017.01.011.

Kadri, S., Long, B. C., Mujacic, I., Zhen, C. J., Wurst, M. N., Sharma, S., McDonald, N., Niu, N., Benhamed, S., & Tuteja, J. H. (2017). Clinical validation of a next-generation sequencing genomic oncology panel via cross-platform benchmarking against established amplicon sequencing assays. *The Journal of Molecular Diagnostics*, *19*(1), 43–56.

Kchouk, M., Gibrat, J.-F., & Elloumi, M. (2017).*Generations of Sequencing Technologies: From First to Next Generation*. 09. https://doi.org/10.4172/0974-8369.1000395.

Kim, B., Won, D., Jang, M., Kim, H., Choi, J. R., Kim, T. I., & Lee, S.-T. (2019). Next-generation sequencing with comprehensive bioinformatics analysis facilitates somatic mosaic APC gene mutation detection in patients with familial adenomatous polyposis. *BMC Medical Genomics*, *12*(1), 103.

Kohlmann, A., Grossmann, V., & Haferlach, T. (2012). Integration of next-generation sequencing into clinical practice: are we there yet? *Seminars in Oncology*, *39*(1), 26–36.

Lee, E. R., Parkin, N., Jennings, C., Brumme, C. J., Enns, E., Casadellà, M., Howison, M., Coetzer, M., Avila-Rios, S., Capina, R., Marinier, E., Van Domselaar, G., Noguera-Julian, M., Kirkby, D., Knaggs, J., Harrigan, R., Quiñones-Mateu, M., Paredes, R., Kantor, R., … Ji, H. (2020). Performance comparison of next generation sequencing analysis pipelines for HIV-1 drug resistance testing. *Scientific Reports*, *10*(1), 1634. https://doi.org/10.1038/s41598-020-58544-z.

Liu, L., Li, Y., Li, S., Hu, N., He, Y., Pong, R., Lin, D., Lu, L., & Law, M. (2012). Comparison of next-generation sequencing systems. *BioMed Research International*, 2012, Article ID 251364.

Mallaret, M., Renaud, M., Redin, C., Drouot, N., Muller, J., Severac, F., Mandel, J. L., Hamza, W., Benhassine, T., & Ali-Pacha, L. (2016). Validation of a clinical practice-based algorithm for the diagnosis of autosomal recessive cerebellar ataxias based on NGS identified cases. *Journal of Neurology*, *263*(7), 1314–1322.

Malone, E. R., Oliva, M., Sabatini, P. J. B., Stockley, T. L., & Siu, L. L. (2020).Molecular profiling for precision cancer therapies. *Genome Medicine*, *12*(1), 8. https://doi.org/10.1186/s13073-019-0703-1.

Maxam, A. M., & Gilbert, W. (1977).A new method for sequencing DNA. *Proceedings of the National Academy of Sciences*, *74*(2), 560–564.

Meldrum, C., Doyle, M. A., & Tothill, R. W. (2011).Next-generation sequencing for cancer diagnostics: a practical perspective. *The Clinical Biochemist. Reviews*, *32*(4), 177–195. PubMed.

Ng, S. B., Buckingham, K. J., Lee, C., Bigham, A. W., Tabor, H. K., Dent, K. M., Huff, C. D., Shannon, P. T., Jabs, E. W., Nickerson, D. A., Shendure, J., & Bamshad, M. J. (2010). Exome sequencing identifies the cause of a mendelian disorder. *Nature Genetics*, *42*(1), 30–35. PubMed. https://doi.org/10.1038/ng.499.

Niguidula, N., Alamillo, C., ShahmirzadiMowlavi, L., Powis, Z., Cohen, J. S., & Farwell Hagman, K. D. (2018). Clinical whole-exome sequencing results impact medical management. *Molecular Genetics & Genomic Medicine*, *6*(6), 1068–1078. PubMed. https://doi.org/10.1002/mgg3.484.

Nomura, J., Hisatsune, A., Miyata, T., & Isohama, Y. (2007).The role of CpG methylation in cell type-specific expression of the aquaporin-5 gene. *Biochemical and Biophysical Research Communications*, *353*(4), 1017–1022.

Qin, D. (2019). Next-generation sequencing and its clinical application. *Cancer Biology & Medicine*, *16*(1), 4–10. PubMed. https://doi.org/10.20892/j.issn.2095-3941.2018.0055.

Raza, K., & Ahmad, S. (2019). Recent advancement in next-generation sequencing techniques and its computational analysis. *International Journal of Bioinformatics Research and Applications*, *15*(3), 191–220.

Raza, K., & Qazi, S. (2019). Nanopore sequencing technology and Internet of living things: a big hope for U-healthcare. In Dey, N., Chaki, J., & Kumar, R. (Eds.), *Sensors for Health Monitoring* (pp. 95–116). Elsevier.

Ronaghi, M. (2001).Pyrosequencing sheds light on DNA sequencing. *Genome Research*, *11*(1), 3–11.

Saito, M., Momma, T., & Kono, K. (2018).Targeted therapy according to next generation sequencing-based panel sequencing. *Fukushima Journal of Medical Science*, *64*(1), 9–14. PubMed. https://doi.org/10.5387/fms.2018-02.

Sanger, F., Air, G. M., Barrell, B. G., Brown, N. L., Coulson, A. R., Fiddes, J. C., Hutchison, C. A., Slocombe, P. M., & Smith, M. (1977). Nucleotide sequence of bacteriophage φX174 DNA. *Nature*, *265*(5596), 687–695. https://doi.org/10.1038/265687a0.

Savelieff, M. (2019).*Next-Generation Sequencing (NGS): Stimulating the Next Generation of Cancer Diagnostics and Treatment*. https://www.technologynetworks.com/diagnostics/articles/next-generation-sequencing-ngs-stimulating-the-next-generation-of-cancer-diagnostics-and-treatment-319649.

Seifi, M., Ghasemi, A., Raeisi, S., & Heidarzadeh, S. (2017).Application of Next-generation Sequencing in Clinical Molecular Diagnostics. *Brazilian Archives of Biology and Technology*, *60*. http://www.scielo.br/scielo.php?script=sci_arttext&pid=S1516-89132017000100303&nrm=iso.

Shen, T., Yeat, N. C., & Lin, J. C.-H. (2015). Clinical applications of next generation sequencing in cancer: from panels, to exomes, to genomes. *Frontiers in Genetics*, *6*, 215.

Siu, L. L., Conley, B. A., Boerner, S., & LoRusso, P. M. (2015). Next-generation sequencing to guide clinical trials. *Clinical Cancer Research : An Official Journal of the American Association for Cancer Research*, *21*(20), 4536–4544. PubMed. https://doi.org/10.1158/1078-0432.CCR-14-3215.

Taylor, T., Lee, E. R., Nykoluk, M., Enns, E., Liang, B., Capina, R., Gauthier, M.-K., Domselaar, G. V., Sandstrom, P., Brooks, J., & Ji, H. (2019). A MiSeq-HyDRA platform for enhanced HIV drug resistance genotyping and surveillance. *Scientific Reports*, *9*(1), 8970–8970. PubMed. https://doi.org/10.1038/s41598-019-45328-3.

Theilade, J., Kanters, J., Henriksen, F. L., Gilså-Hansen, M., Svendsen, J. H., Eschen, O., Toft, E., Reimers, J. I., Tybjærg-Hansen, A., & Christiansen, M. (2013).Cascade screening in families with inherited cardiac diseases driven by cardiologists: Feasibility and nationwide outcome in long QT syndrome. *Cardiology*, *126*(2), 131–137.

Vrancken, B., Lequime, S., Theys, K., & Lemey, P. (2010).Covering all bases in HIV research: unveiling a hidden world of viral evolution. *AIDS Reviews*, *12*(2), 89–102.

Whole Genome Association Studies. (2011). https://www.genome.gov/17516714/2006-release-about-whole-genome-association-studies.

3 Clinical Applications of Next-Generation Sequence Analysis in Acute Myelogenous Leukemia

Fatima Nazish Khan
Department of Computer Science, Jamia Millia Islamia

Shaban Ahmad
Agricultural Knowledge Management Unit,
ICAR-Indian Agricultural Research Institute

Khalid Raza
Department of Computer Science, Jamia Millia Islamia

CONTENTS

3.1 INTRODUCTION

Acute myelogenous leukemia (AML), also known as acute myeloid leukemia, acute granulocytic leukemia, acute nonlymphocytic leukemia, and acute myeloblastic leukemia, is the second most common hematologic malignancy distinguished by increased growth and collection of undifferentiated unformed myeloid progenitor blood cells in the bone marrow, which leads to an accumulation of immature cells causing impaired hematopoiesis and failure of the bone marrow (Papaemmanuil et al. 2016). In simpler words, it is a clonal disorder originated from genetic abnormalities in common myeloid hematopoietic precursor cells, which leads to alteration in differentiation, self-renovation, and unlimited growth of cells. It is considered as the unexceptional acute leukemia in patients of older age (>65 years) with an outcome of up to 17 per 100,000 persons per year (Papaemmanuil et al. 2016). In the past four decades, induction of chemotherapy using cytarabine and daunorubicin has improved clinical results in young patients. However, the survival rate for older patients was not satisfactory (Burnett et al. 2011) though other factors such as old age, poor health status, comorbidity, and genetic abnormalities such as mutations were also related to the results (Appelbaum et al. 2006). Hence, the ailment is considered as a very heterogeneous system from the evidence of aberrations in chromosomal rearrangement by cytogenetics and molecular genetics methods, and it is the best suited for moving ahead into the biology of leukemia. In the past decade, there has been a huge increment in our mutational understanding of AML (Papaemmanuil et al. 2016; Patel et al. 2012) with the advancement of sequencing techniques. The genome of AML is identified as the least difficult malignancy genome, which makes it amiable to the clinical utilization of next-generation sequencing (NGS).

Previously, the large size (approx. 3 billion base pairs) and complex structure of the human genome made it impractical for sequencing of cancer genomes in a short period of time (Papaemmanuil et al. 2016). Two significant advances assisted to overcome these obstructions. First, the Human Genome Project (HGP) in 2001 has generated the draft sequence and a master plan of the whole human genome. Second, the advancement of DNA sequencing techniques progressively decreases the rate and time needed for sequencing genomes (Figure 3.1). It has been notified that initially the HGP spent more than a decade and billions of dollars to sequence the human genome. Nowadays, it only takes 5–6 weeks and $20,000 per human genome and around $40,000 for paired malignant/commonly found genomes. Consequently, as we quickly move toward the genome sequencing of patients with malignancy, it will be useful in clinical applications.

Cancer, in its many forms and classification, was the cause of 8.2 million deaths in 2012. Therapeutic improvements in the sequencing techniques of DNA have paved a new way to transform our perception of this profoundly perplexing and several categories of ailments (DeVita and Rosenberg 2012). First, the translocation (t(9;22)) associated with chronic myeloid leukemia was identified by Hungerford and Nowell (Papaemmanuil et al. 2016) through a large amount of data which show the information of mutation and karyotyping of various genes and further give us useful diagnostic, therapeutic, and treatment-related recommendations for AML patients. However, previous techniques have the limitation of expensiveness to identify chromosomal

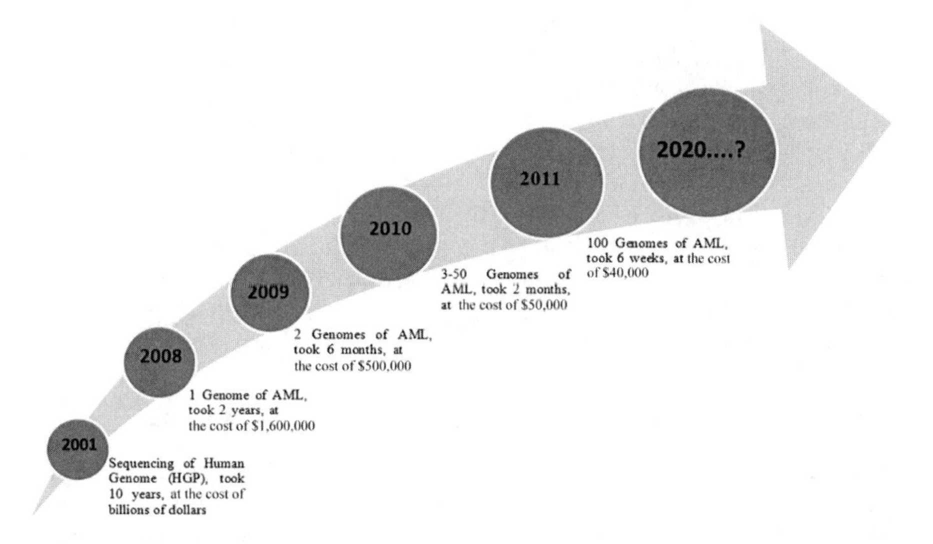

FIGURE 3.1 Timeline of the sequencing of AML genomes, and estimated rate and time taken for sequencing.

aberrations and genotyping of the patients' genome, and due to the increase of genetic abnormalities that occurred in AML patients, this approach is no longer in use. The improvement in variant recognition was initially accomplished and analyzed by array-based technologies such as single-nucleotide polymorphism (SNP) and comparative genome hybridization (CGH). In the past decade, NGS techniques have been recognized as the technology that permits accurate, fast, and massive parallel sequencing of various genes, complete genome, and even exomes within a day and at low cost, which led to the improvement of genomic research and applications. This has brought to light the complexities in identification of new and corresponding mutated genes, evaluation of their functional roles, prognostic stratification, classification, and treatment and response assessment of AML and co-occurrence patterns. The advent use of whole-genome sequencing (WGS) paved a new way for genome-wide coverage, profiling of gene expression, and analysis of nucleotide polymorphism and CGH in AML patients to classify AML biologically.

Although the pathogenesis of AML starts from mutations in many genes, only little of them are clinically relevant. In 2008, the classification system given by the World Health Organization (WHO) characterized AML based on recurrent genetic mutations in two of the oncogenes, namely, *NPM1* and *CEBPA* (Hasserjian 2013). Thus, the classification of WHO in 2016 has given the significance of molecular aberrations by remarking biallelic *CEBPA* and *NPM1* as an isolated system and identifying *BCR-ABL* AML and *RUNX1* as advanced categories. The other recognized mutations of AML are those occur in *ASXL1*, *CEBPA*, *DNMT3A*, *FLT3*, *IDH1/2*, *KIT*, *KMT2A*, *NPM1*, *RUNX1*, *TET2*, *TP53*, and *WT1* genes. The recommendations of European Leukemia Net (ELN) published in 2017 have identified abnormalities in *ASXL1*, *TP53*, *NPM1*, *TP53*, biallelic *CEBPA*, *ITD*, and *RUNX1* to enhance the effect of patients' results for the diagnosis and management of AML (Döhner et al. 2017), and thereafter the AML patients are classified into three groups, namely, good, intermediate, and poor risk. A patient's cytogenetic profile with favorable outcomes includes translocations in *PML-RARA*, *RUNX1-RUNX1T*, and *CBFB-MYH11* genes. Patients of acute promyelocytic leukemia (APL) with mutations in *PML-RARA* genes are treated with all-trans-retinoic acid (ATRA) and arsenic trioxide, or ATRA and anthracycline. However, patients of core binding factor (CBF) leukemias with mutations in *CBFB-MYH11* and *RUNX1-RUNX1T* genes are treated with intensive chemotherapy with cytarabine having favorable prognosis.

This chapter delivers the idea for two purposes. First, it will focus on how the ongoing effect of NGS is being utilized to study and help in comprehension and analysis of cancer genomes, particularly in AML. Second, it will highlight some of the challenges and future trends of NGS technology in the field of cancer genomics.

3.2 BACKGROUND AND HISTORY OF SEQUENCING

DNA sequencing was initiated in the early 1970s with the help of methods utilized by four scientists Sanger-Coulson and Maxam-Gilbert. In the Maxam–Gilbert approach, DNA fragments were labeled by radioactive probes at the terminals, and then the fragments were cleaved using specific chemicals. Subsequently, the fragments were separated by gel electrophoresis and then visualized by the autoradiographic film.

The Sanger–Coulson method was based on the utilization of a specified primer to initiate the amplification process at a specific location on the DNA template. This method recruits a dideoxy terminator for DNA molecule which helps in base-specific termination of the synthesized DNA, attaching labels on each nucleotide and accompanying the segregation of chain termination outcomes by using capillary electrophoresis (Dovichi 1997; Sanger et al. 1977; Raza and Ahmed 2019). Hence, the dideoxy chain termination method of Sanger was adopted as a significant method for DNA sequencing because no or fewer chemicals are required, and it is highly efficient and is suitable for particular gene sequencing which can precisely read approximately 900 bp. Although the Sanger sequencing method is widely used around the world in different labs, this method is not able to fulfill the requirements of the research scholars and scientists because of many drawbacks like low speed, less throughput, high cost, and time-consuming. To improve the quality of information and to overcome these limitations, a high-throughput sequencing (HTS) technology was needed. With the discovery of a HTS technology known as NGS technology announced by Helicos Heliscope™ and Pacific Biosciences in 2005, a large amount of whole-genome DNA could be quickly sequenced with high accuracy in a cost-effective manner (Biesecker et al. 2012). NGS, also known as deep sequencing technology, is a massively parallel sequencing technique that results in several small sequences amplifying millions of DNA fragments on an array to know the sequence that will be received in one round (as shown in Figure 3.2) (Morozova and Marra 2008). In comparison with the

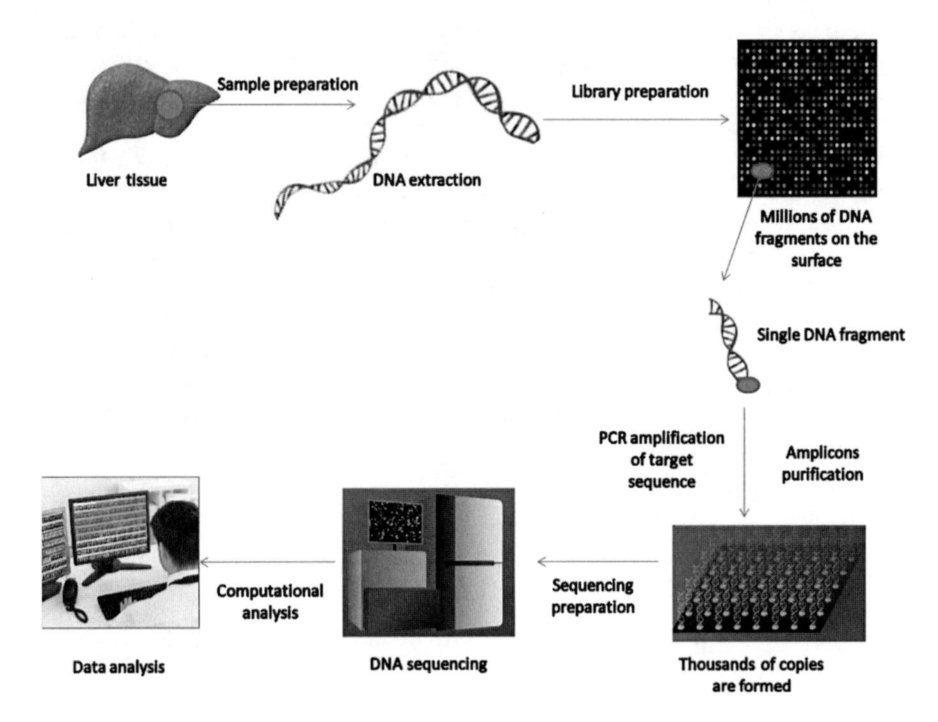

FIGURE 3.2 Steps showing the sequencing of genomic DNA.

traditional methods of sequencing, this technology is progressively revolutionized the fields of biomedicine, genomic research, and molecular biology. Moreover, it is much faster than the traditional sequencing done by Sanger because of merging the features of reaction between chemicals and analyzing signals with each other and reducing the requirement for the methods of fragment cloning as utilized in the traditional methods. The main NGS technologies that are mostly in use today are Illumina, Solexa, Roche, LS454, ABI-SOLiD, and Helicos, and most recently, Oxford Nanopore Technology based on electronics helps in the measurement of electrical current changes as molecules pass throughout nanopores (including GridION, PromethION, SmidgION sequencers, and portable pocket-size MinION sequencing device). These methods are now globally used in clinics for many purposes such as molecular diagnostics, recognition of allelic variants, SNPs, Mendelian and complex genetic disorders, understanding RNA structure, transcriptomics, differentiation of different types of cancers, detection of pathogens, epigenetic changes, prenatal screening, and personalized medicine (Koboldt et al. 2013).

3.3 ACUTE MYELOGENOUS LEUKEMIA

3.3.1 AML and Its Morphology

The AML, commonly called as acute myeloid leukemia, is a class of heterogeneous disease caused in the bone marrow and peripheral blood cells due to the clonal enhancement of myeloid progenitors. Previously, AML could not be cured due to limitations in the availability of resources, but the advancement of technologies has rehabilitated 35%–40% of patients below 60 years. However, it still remains challenging to cure patients older than 60 years. Many studies have shown that this disease arises from the mutations recurrently accumulated in hematopoietic stem cells with age. As the NGS technologies develop, primary and relapsed tumors have been identified with their therapeutic approaches (Ding et al. 2012). It has been characterized morphologically that the progenitors of AML have varying sizes and shapes from as small as lymphocytes to as large as monocytes, with a large-size nucleus usually containing several nucleoli. The precursors of AML are present on fit and fine undeveloped myeloid cells such as common differentiation (CD) markers including CD-13, CD-33, and CD-34 (Campos et al. 1989) and are usually expressed based on morphological subclassification of AML and differentiation of immature cells. Additionally, it can also co-express T-cell and B-cell antigens such as terminal deoxynucleotidyl transferase, human leukocyte antigen–antigen D related, CD-7, and CD-19.

3.3.2 Classification Systems for AML

Many different classification systems have been discovered for AML in the past years according to their morphology, etiology, phenotypic immunity, and genetics study, and in the early 1970s, it was classified based on the French-American-British system of classification through the criteria of morphological appearances and phenotypic and cytochemical immunity, to elucidate eight AML subtypes such as FAB M0, M1,

M2, M3, M4, M5, M6, and M7. But the old French-American-British system of AML classification is now replaced by the classification of WHO. WHO classification has been renewed in 2008 and defines seven subtypes of AML such as AML having myelodysplasia-related changes; AML having repeated genetic aberrations such as aberrations in *RUNX1-RUNX1T1* translocation t(8;21)(q22;q22), *CBFB-MYH11* inversion Inv(16)(p13.1q22), and translocation t(16;16)(p13.1;q22), *PML-RARA* translocation t(15,17)(q22;q12), *MLL* 11q23, and with mutated genes such as *NPM1* and *CEBPA*; NOS-AML; therapy-related myeloid neoplasms; the proliferation of myeloid based on Down's syndrome; myeloid sarcoma; and blastic plasmacytoid neoplasm in dendritic cells (Vardiman et al. 2009). According to the classified system of etiology, AML can be categorized as secondary AML, therapy-related AML, and de novo AML (Lindsley et al. 2015).

3.3.3 Cytogenetics

Approximately 50%–60% of all young primary patients of AML are represented with chromosomal aberrations such as indels or translocations that increase the ailments of AML. Cytogenetic mutations such as translocation of (8;21)(q22;q22), (15;17)(q22;q12), and inversions of (16)(p13.1;q22) are characterized by better and longer overall remission and survival, but the mutations in chromosomes 5, 7, and 11q23 are recognized by shorter overall survival and poor therapy responses. About 50% of the AML cases are categorized as cytogenetically normal AML or CN-AML (Gaidzik and Döhner 2008), which is characterized as an intermediary risk of relapse in terms of clinical outcomes.

3.3.4 Molecular Abnormalities in AML

In the past years, many studies have identified that there are specific gene mutations that exist in the genome of AML patients, having effect on the treatment and diagnosis of CN-AML (Lindsley et al. 2015; Marcucci et al. 2011). With the advancement of NGS techniques, the genetic idea of CN-AML patients has been identified with 13 mutations; out of these, 8 are randomly identified passenger mutations and 5 are recurrently driver mutations. Most of the relevant AML mutations have been identified such as *RUNX1*, *CBFB*, *PML-RARA*, *MECOM*, *MLLT3*, *GATA2*, *CEPBA* (biallelic), *NPM1*, *ASXL1*, *KMT2A*, *TP53*, *IDH*, *FLT3*, *DNMT3A*, *TET2*, *SRSF2*, *TP53*, *c-KIT*, *SF3B1*, and *DEK*.

3.3.5 Prognosis/Risk Stratification

Instead of chromosomal and molecular aberrations, age factor and performance status are important strategies to predict the outcomes in AML patients. Therefore, the classification of the ELN was established in 2010 to adopt the risk stratification in adult patients with AML based on cytogenetic and molecular mutations (Döhner et al. 2017), and hence, patients are categorized as favorable, intermediate 1, intermediate 2, and adverse (Table 3.1).

TABLE 3.1
ELN Classification of Genetic Abnormalities

Risk Category	Abnormalities Related to Genetics
Low or favorable risk	t(8;21)(q22;q22); RUNX1-RUNX1T1
	t(16;16)(p13.1;q22); CBFB-MYH11
	NPM1 (mutated) without FLT3-ITD
	Biallelic CEBPA (mutated)
Intermediary-1 risk	NPM1 (mutated) and FLT3-ITD
	Wild-NPM1 and FLT3-ITD
	Wild-NPM1 without FLT3-ITD
Intermediate-2 risk	t(9;11)(p22;q23); MLLT3-KMT2A
	Cytogenetic difficulties are not considered as favorable
	Cytogenetic difficulties are not considered as adverse
High or adverse risk	inv(3)(q21q26.2); GATA2-MECO (EVI1)
	t(6;9)(p23;q34); DEK-NUP214
	t(v;11)(v;q23); rearranged KMT2A
	–5 or del(5q); –7; abnl(17p);

3.3.6 Therapeutics for AML Patients

Since 1970, the intensive induction chemotherapy has been utilized as a therapeutic for the AML patients, for example, the utilization of anthracycline and cytarabine, which is suitable for younger individuals aged less than 60 years. Additionally, the dose of either daunorubicin or idarubicin with cytarabine infusion for 7 days aims at achieving morphologic complete remission (CR) and to treat AML patients (Cheson et al. 2003). Some patients have a mutation in the *FLT3* gene and have been treated with an *FLT3* inhibitor like midostaurin along with standard induction therapy. Intensive induction therapy is sometimes not suitable for older patients, and hence, they are treated with hypomethylating drugs such as decitabine and azacitidine (Klepin 2014; Quintás-Cardama et al. 2012), which are used for the treatment of myelodysplasia. Patients that have APL are given the treatment of ATRA before the confirmation of diagnosis, which reduced the risk of disseminated intravascular coagulation (DIC) development, APL-induced coagulopathy, and even mortality.

3.3.6.1 Strategies of Consolidation

Consolidation is acknowledged as a therapeutic post-induction method that is applied to patients for slow prevention and minimal residual disease (MRD) in the bone marrow. With the help of NGS technologies or real-time PCR, we can assess MRD, which can help in tracking the response to treatment and is the best method for prediction of relapse (Grimwade and Freeman 2014). Instead of this exact information present, the AML heterogeneity will lead to promotion of mutational clones which may lead to difficulties in identification and characterization of leukemia's risk development. Generally, the two most important strategies that can be applied for consolidation are hematopoietic stem cell transplantation and chemotherapy of targeted agents (Estey and Döhner 2006), which can be used alone or in combination depending on the fitness of the patient, the type of leukemia, and the presence of stem cell donors.

The post-induction chemotherapeutic method used a minimum dose of cytarabine, that is, 1.5 g/m^2 two times daily on alternative days (i.e., 1, 3, and 5) within three to four cycles to increase the remission rate and improve the survival rate in young adult patients aged less than 60 years, followed by chemotherapy and then by transplantation. Furthermore, Burnett et al. (2013) in their studies challenged giving a higher dose, that is, 3 g/m^2 instead of 1.5 g/m^2, to provide the similar results of cytarabine in the same age group, for the treatment of patients with *CBF* AML and *NPM1*-mutated AML. The minimum or higher doses of cytarabine gave no benefit to older patients aged more than 60 years (Schiffer 2014). In prognostic groups including intermediate- or high-risk fit patients achieved, the transplantation of allogeneic hematopoietic stem cell was effectively used for long-term therapeutics for AML patients and has cured 50%–60% of the patients with initial CR. However, most of the patients are not suitable for transplantation due to several comorbidities or shortage of eligible donors. Hence, patients are given post-induction chemotherapy to perpetuate CR and help decrease the load of leukemia. Sometimes the consolidation therapy would lead to the risk of mortality and morbidity, and hence, it might be suitable to move forward with transplantation rather than consolidation. Most of the recent studies gave the evidence that age is not the individual criterion for transplanting, but the decision of eligibility has been taken up based on the performance of pre-transplantation comorbidities and what is the stage of remission by using Hematopoietic Cell Transplantation Comorbidity Index (HCT-CI) tool (Sorror et al. 2005). The methods that include effective transplantation in all age groups are supportive care, enhancement of donor options such as cord grafts and haploidentical donors, and reduction in the intensity preparation regimens for HCT.

3.3.7 Novel Targets for AML

3.3.7.1 FLT3 Inhibitors

Inhibitors of *FLT3* are the first-generation drugs which inhibits kinase, such as *quizartinib*, *crenolanib*, *lestaurtinib*, *cytarabine*, *sunitinib*, *midostaurin*, *sorafenib*, *midostantrum*, *5-azacitadine*, *KW2449*, and *tandutinib*, and presents with anti-leukemic activity which enhances the toxicity and reduces blood and bone marrow blasts (Sudhindra and Smith 2014). These can be utilized alone or in combination depending on the fitness of the patient and the type of leukemia.

3.3.7.2 Isocitrate Dehydrogenase Inhibitors

In two different phase I clinical trials, the isocitrate dehydrogenase (IDH) inhibitors AG120 and AG221 have played a significant role in patients with AML (Hansen et al. 2014) having responses of 40% and 31%, respectively. The time period of response is very long in both types of IDH: 15 months for AG221 and 11 months for AG120.

3.3.7.3 Nuclear Exporter Inhibitors

These are basically reversible inhibitors of anti-leukemia CRM1, also called XPO1, which help in the inactivation and export of many tumor suppressors (Fukuda et al. 1997). In hematological cancers, such as AML, and in solid tumors, CRM1 has been upregulated.

3.3.7.4 Immunological Therapies

Immunological and antibody-based therapeutics are also used in the diagnosis of leukemia and are in the phase of development for AML. These days, the monoclonal antibody such as CD33 and bispecific antibody such as AMG330 are being utilized (Castaigne et al. 2012). T cells such as chimerical antigen receptor-transduced T cells are also developed as the target for antigenic receptor target, for example, CD123.

3.4 NEXT-GENERATION SEQUENCING TECHNOLOGIES

3.4.1 Terminology and Its Recent Developments

In the last 10 years, the traditionally utilized Sanger sequencing approach has been overhauled by high-throughput techniques known as NGS technology, which is utilized to illustrate several modern massively parallel sequencing techniques that produce numerous sequencing reads by amplifying and immobilizing millions of DNA fragments and in parallel to identify and analyze the sequencing reads (as shown previously in Figure 3.2) (Morozova and Marra 2008). This massively parallel sequencing approach sequenced a large amount of genomic DNA quickly with high accuracy and cost effectively. With the advancement of NGS technology, we can significantly make our perception of cancer genomes. The various recently developed tools of NGS such as Mi-Seq (sequencing by synthesis) and Ion Torrent PGM (semiconductor sequencing) have been utilized in clinical settings and research laboratories. These instruments are very small, highly accurate, and exhibit reduction in time and cost. These are high-throughput machines that give a large number of reads and are best for analysis in the applications of treatment and diagnosis (Ley et al. 2008).

3.4.2 How NGS Is Going from Research to Clinical Settings: An Overview

While NGS technologies are very costly and time consuming, there recently has been a reduction in costs for sequencing a whole human genome—$100 million USD in 2001 to approximately $1,000 in 2015. Similarly, there is a huge reduction in the price of sequencing the reads from approximately $10,000 in 2001 to less than 10 cents in 2011 (as shown previously in Figure 3.1). There are many platforms of HTS technology available in the market, such as Illumina (San Diego, CA) platforms based on the methods of sequencing by synthesis (including miniSeq, iSeq100, nextSeq, miSeq, NovaSeq, HiSeqX Ten, and HiSeq2500 sequencers), Thermo Fisher Scientific (Waltham, MA) platforms which use semiconductors to measure the alterations in pH results from release of hydrogen ions with the addition of nucleic acids (including Ion PGM System, Ion Proton System,

Ion S5 XL System, Ion S5 System, Ion Gene Studio S5 System, and HID Gene Studio S5 System), Pacific Bioscience based on one molecule real-time technology (including the PacBio System platform), and Oxford Nanopore Technology based on electronics that help in the measurement of electrical current changes as molecules pass through nanopores (including GridION, PromethION, SmidgION sequencers, and portable pocket-size MinION sequencing device). There is a major difference in costs in all types of sequencers (WGS, whole-exome sequencing (WES), and targeted sequencing) depending on read size, sample number, the quality control methods applied, the number of megabases in the panel, and the sequencing platform used. Nowadays, different myeloid NGS gene panels are available commercially covering many AML genes including AmpliSeq® Myeloid Sequencing Panel, Leuko-Vantage Myeloid Neoplasm Mutation Panel, Human Myeloid Neoplasms Panel, and SureSeq myPanel™ NGS Custom AML.

3.4.3 Clinical Applications of NGS Technologies in Cancer Research

With the advancement of the NGS techniques, our understanding to work on malignant genomes has been revolutionized. These sequencing technologies generate millions of sequencing reads in parallel of ~50–100 nucleotides in length, followed by the mapping of these sequences on the reference human genome to gather the information of the malignant genome. Usually in the study of cancer, there is the need to do the sequencing of both normal and tumor tissues. In 2001, when the HGP was finished, first cancer exome was sequenced having only 1%–2% of the genome that has been translated. Soon after that, in 2003, Ley et al. proposed an experiment and used the NGS technology to study the 140 exomes of human AML cells, which identified 13 abnormalities of the pathogenesis of AMP. Currently, there are different methods to study cancer genomes with the use of NGS techniques.

3.4.3.1 Whole-Genome Sequencing

Whole-genome sequencing (WGS) is characterized by taking and sequencing the entire genome of any individual. It gives the idea of the entire human genome which comprises 39,109 (3 Gbases) genes and includes both the coding and noncoding regions. WGS is based on NGS techniques that enable the characterization of genomic aberrations such as indels, point mutations, copy number variations (CNVs), inversions, translocations, amplifications, and uniparental disomy in the entire genome of the organism. The major drawbacks of WGS are its cost, complicated analysis, and longer turnaround times; instead, it is the best and dominant method for the detection of mutations. Initially, in 2008, the genome of cancer was sequenced by Ley et al. on leukemia cells, and the experiment was done on one patient and the cells present in the skin of the patient were considered as a template and detected 10 somatic mutations; that is, two genes (*FLT3* and *NPM1*) are known to contribute in tumor progression and eight have unknown functions. After some time, they also sequenced another patient of AML having standard cytogenetics and identified 64 somatic mutations which present in the consensus sequence of the genome as well as in the coding sequences of genes (Welch et al. 2011), followed by further validation in a large study on 187 samples of AML having aberrations in most of the genes such as *NPM1*,

NRAS, and *IDH1*. By resequencing the samples of AML from the initial study, Ley et al. in 2010 utilized the method of paired-end deep sequencing which identifies the recurring of somatic mutation in the *DNMT3A* in 22% (62/281) of patients. Similarly, Welch et al. also used WGS for the treatment of an AML case from their institute for the transplantation of stem cells. The patient chosen had a complex cytogenetic description with no characterization of *PML-RARA* fusion but had contained APL. After the performance of WGS, he identified fusion genes of *PML-RARA*, *RARA-LOXL1*, and *LOXL1-PML*. Ultimately, it leads to extended remission when the victim of AML was treated as APL. Welch et al. in their other cohort of WGS identified initiating mutation in CN-AML patients and thereafter concluded that approximately the same number of alterations was found in both AML groups. After some time, the clonal evolution of mutations was reported by Ding et al. in acute and relapsed myeloid leukemia (Burrell and Swanton 2014). The study of the Cancer Genome Atlas Research Network also analyzed 50 de novo AML cases by using WGS in a combination of methylome and whole-exome analysis and found that the mutation in AML genomes is less as compared to other cancers (Ding et al. 2012).

3.4.3.2 Transcriptome Sequencing

Researchers use this method for sequencing and analyzing all genes that are transcribed, including both those code for proteins and noncoding RNAs, commonly known as the transcriptome. Major advantages include identification of expressed fusion transcripts, gene expression-level information, and posttranscriptional changes such as SNVs and alternative splicing. This technique cannot identify and characterize mutations in the noncoding portions of the genome; that is, it will not identify mutations in genes that accelerate RNA turnover such as nonsense mutations or frameshift mutations and the mutations affected by the loosing of gene copies. The recent advancement of transcriptome sequencing is RNA, called RNA-Seq, engaged in the selection of poly-A, synthesis of cDNA by reverse transcription of RNA, disintegration of genes, and finally accomplished by attaching the primers (McGettigan 2013; Jabeen et al. 2018). In a category of 45 patients of AML, Wen et al. characterized and identified novel fusion transcript by utilizing a paired-end library of RNA-Seq data, which include the cases of 29 CN-AML, 8 with abnormal karyotype, and another 8 without any karyotype information (Wen et al. 2012). The identification and characterization of fusions in patients of AML with abnormal karyotype have shown *RUNX-RUNX1T1*, *MLL-MLLT1*, and *MLL-MLLT3* with the help of the RNA-Seq method. In all AML cases of 134 identified fusions, CN-AML usually contains seven transcript fusions. Currently, with the help of transcriptome sequencing, two novel fusion transcripts are identified in CN-AML patients.

3.4.3.3 Whole-Exome Sequencing

The exome is described as the protein-coding portion of the genomic DNA that may be present as only 1%–2% of the whole genome. Whole-exome sequencing (WES) is a method that specifically analyzes and validates only the part of the genome which codes for proteins together with micro-RNAs and other noncoding RNAs. Because only 1%–2% of the genome is analyzed through WES, it has major advantages of cost reduction and relatively deep sequence coverage. However, it has some limitations:

it is unable to detect any mutations in regions away from the exome, that is, above 98% of the genome and structural variants such as chromosomal translocations typically present in intronic regions, larger insertions and deletions. Similar to WGS, it also needs paired-end samples to differentiate mutations of polymorphism in cancer patients. In the case of AML, it has applications in clinical trials when any clinical response is related to mutation. With the help of exome sequencing, Yan et al. (2011) recognized abnormalities in 14 genes of 9 AML-M5 patients in his study. They found that out of 14 genes, 6 have mutations in cancers or other diseases because of 3 *DNMT3A* variants. MLL abnormalities were also identified, which include MLL translocations, *FLT3* internal tandem duplications, and *NPM1* and *NRAS* mutations (Yan et al. 2011). Using WES, Grief et al. sequenced *PML-RARA* in three APL positive patients in states of leukemic and remission, and they identified that most of the genes increase the risk of the pathogenesis of the present ailments. Similarly, the study by Grossman on CN-AML patients by using WES showed that there is no disease found in any of the genes. In one of the studies, mutations in 11 genes were found for AML pathogenesis (Grossmann et al. 2011). In another study using exome sequencing, Grief et al. identified that two out of five biallelic *CEBPA* AML patients contain novel abnormalities of *GATA2*. With the utilization of exome sequencing in CBF leukemia, Opatz detected novel *N676K* mutation in *FLT3*, which helps in the detection of genetic aberrations in the genome.

3.4.3.4 Targeted Next-Generation Sequencing

Targeted NGS is a technique that sequences only specified genes that are mostly used by clinical laboratories for the molecular diagnostic profiling of AML patients. The vital advantages of targeted NGS are typically the low cost, large coverage depths, and rapid turnaround times, which enable the detection and characterization of allelic variants as small as 2%. Drawbacks include problems in detecting variations in genes and structures such as CNVs and SNPs. Targeted NGS includes several different methods such as multiplex PCR, hybrid capture, and ligation-mediated methods, which select the genome including protein-coding and noncoding regions and identify chromosomal translocations (Duncavage et al. 2012). With the help of deep sequencing, Conte et al. used targeted NGS to identify 24 gene mutations in CN-AML patients such as *NPM1*, *FLT3*, *CEBPA*, *TET2*, *DNMT3A*, *KRAS*, *IDH1*, *IDH2*, and *WT1*. By using this technique in cell lines of AML, Duncavage et al. searched for a way that identifies and characterizes mutations in a gene such as indels, SNVs, and translocations. Recently, the cohort of Kihara uses this technique on 51 genes and found that 44 genes are mutated with 505 mutations such as in *CEBPA*, *FLT3*, *DNMT3A*, *NPM1*, *MLL-PTD*, *TP53*, and *KIT* in 5%–10% of the patients of AML (Kihara et al. 2014). As time progresses, with the advancement in targeted NGS, most of the clinical settings are benefited in the identification and characterization of abnormalities in genes, require less processing time, and lead to risk stratification and better diagnosis of AML patients.

3.4.3.5 Other Applications of NGS

To identify and distinguish the malignant genes from epigenetic changes, NGS applications are preferred, especially in the study of AML. Most of the genes that regulate DNA methylation such as *DNMT3A* are simultaneously changed in the patients of AML. Therefore, NGS techniques are helpful in the genome-wide mapping of methylated DNA with the best resolution (Lister et al. 2009) and in identification of the structure of chromosomes and further chromatin structure by mapping and posttranslational modifications.

3.5 NGS ANALYSIS GUIDELINES FOR THE DIAGNOSIS OF AML

Recently, there is the recommendation and guidelines published by the ELN to test genetic abnormalities in all patients having AML, which are globally accepted by people in common practices, in clinical settings, and by marketing agencies. The first edition of ELN came in 2010, having most of the outcomes from standard cytogenetic and aberrations in most of the genes such as *FLT3, NPM1*, and *CEBPA*. It categorizes AML patients into four categories, namely, favorable, intermediate-1, intermediate-2, and adverse disease. The updated version of ELN published in 2017 divides AML patients into three categories, namely, favorable, intermediary, and adverse (as shown in Table 3.1), based on the outcomes of standard cytogenetic and gene abnormalities found in a minimum of six genes such as biallelic *CEBPA*, *FLT3*, *NPM1*, *ASXL1*, *TP53*, and *RUNX1*. Similarly, there are the guidelines of the National Comprehensive Cancer Network which recommend testing for genera arrangements the same as ELN recommendations and conventional cytogenetics, which include three more gene mutations to ELN recommendations, and include a screening for nine aberrations in genes, viz. *CEBPA*, *NPM1*, *IDH1*, *IDH2*, *FLT3*, *RUNX1*, *ASXL1*, *TP53*, and *c-KIT* (Gerstung et al. 2017), and gene rearrangements screening, for example, in *CBFB-MYH11*, *BCR-ABL1*, *PML-RARA*, and *RUNX1-RUNX1T1* genes. According to the guidelines of both the institutions, mutational testing has been utilized as a minimum for day-to-day clinical practices to know accurately what are genomic risk stratifications and whether targeted therapy was used appropriately. They both utilize multiple NGS gene panels to further take out the data of treatment and prognosis of AML patients.

Due to the active research area of genomics and molecular characterization of AML, further modifications are needed in the classification system in near future. The NGS panels used for the clinical purpose might help in the identification of mutations that are diagnostic, prognostic, and predictive. At this point, there is no idea of how many genes or what compositions of genes are to be included in a myeloid NGS panel. Most of the genes are analyzed in AML, and their frequency of mutations is published in the major research of AML characterized by NGS techniques. With the advancement of the biological view of AML pathogenesis, eight new drugs for AML have been approved in April 2017, which enhances our prognostic understanding of the struggle against AML. The updated version of ELN recommendations in 2017 has given the place for the negativity of MRD in response to AML as a new category, by assessing two methods, namely, multicolor flow cytometry and RT-qPCR. A 2017 ELN classification and the cohort of Papaemmanuil et al. (2016) suggest risk stratification of patients of NGS panel to novel prognostic and conclude that 46%, 28%, and 26% of patients are in the categories of favorable, intermediate, and unfavorable risk, respectively.

Using NGS in biological and clinical practices such as identification of genetic aberrations in AML patients remains challenging. Additionally, it is much difficult to utilize NGS testing times with TAT as published in the current guidelines of ELN, and hence, there is a need for conventional molecular testing for the correct diagnosis of myeloid neoplasm (MN).

3.6 HOW AML GENES IMPLICATE MUTATIONS? A PROGNOSTIC AND THERAPEUTIC OVERVIEW

3.6.1 Understanding Cytogenetic and Molecular Genetic Mutations in AML

Various molecular genetic aberrations and cytogenetic mutations have a high impact on therapy-related outcomes in AML patients. For AML classification, the cytogenetic mutations are considered as the best suited prognostic indicators, genetic biomarkers, and treatment outcome present at diagnosis. When inversion of inv(16) or translocation of t(8;21) is present and acknowledged as CBF-AML, intensive chemotherapy shows the best result compared to CN-AML (Dombret et al. 2009). The AML patients that have single autosomal monosomies and complex karyotype AML (CK-AML) have shown poor prognostic implications, whereas the AML patients that have several numbers of autosomal monosomies have poor outcomes of prognostic implications. This poor prognosis of AML can be stratified as when molecular abnormalities are added (Rücker et al. 2012). Similarly, the *KIT* mutation that exists in CBF-AML also shows unfavorable results. Recent mutations in *NPM1*, *FLT3*, and *CEBPA* are considered as useful strategies in prognostic subclassification of CN-AML. AML contains approximately 20% of *FLT3* internal tandem duplication (ITD) and 28%–34% present in CN-AML (Döhner and Gaidzik 2011). The favorable results are seen when there is no *FLT3-ITD* in patients with mutated *biCEBPA* or *NPM1* gene. As with the rise in NGS technologies, most of the mutations have been described by many authors (Table 3.2).

TABLE 3.2
Genetic Abnormalities Found in AML by Different NGS Approaches

AML Type	NGS Techniques	Mutation	Frequency of Identified Recurrent Mutations
AML-M1	Transcriptome seq	Missense mutation stop mutation	TLE4 SHKBP1 RUNX1
AML-M5	Exome seq	Missense translocation	DNMT3A (20.5) GATA2 (3.6) MLL(19.6)
BiCEBPA+ AML	Exome seq	Missense mutation	GATA2 (39.4)
CBF leukemia	Exome seq	Missense mutation	FLT3-N676K (6)
CN-AML	WGS	Frame shift insertion missense	NPM1 (23.9) NRAS (9.3) IDH1 (8.5)

(Continued)

TABLE 3.2 (*Continued*)
Genetic Abnormalities Found in AML by Different NGS Approaches

AML Type	NGS Techniques	Mutation	Frequency of Identified Recurrent Mutations
CN-AML	WGS	Non-synonymous SNV	DNMT3A (22.1)
CN-AML	WGS	Non-synonymous SNVs	TTN DNMT3A NPM1 FLT3 WT1 RUNX1 IDH2
CN-AML	Targeted DNA capture	Single-base substitution indels	FLT3 NPM1 CEBPA DNMT3A TET2 IDH1 IDH2 WT1 RAS
CN-AML	Transcriptome seq	–	CIITA-DEXI fusion transcript (14/29)
CN-AML (NPM-FLT-CEBPA-MLL-)	Exome seq	Disruptive mutation	BCOR (17.1)
CN-AML/M1	WGS	Indels	FLT3 (27.6) NPM1 (23.9)
De-novo AML	WGS & Exome seq	Non-synonymous SNVs	NPM1 FLT3 DNMT3A IDH1 IDH2 NRAS RUNX1 TET2
Pediatric CN-AML	Transcriptome seq	–	CBFA2T3-GLIS2 fusion transcript (8.4)
Pediatric CN-AML CBFA2T3-GLIS2 +	Transcriptome seq	–	DHH-RHEBL1 fusion transcript (40)

The family of DNA methylation enzymes contains a member called *DNMT3A* having three members of mammalian class *DNMT1*, *DNMT3A*, and *DNMT3B*, in which *DNMT1* helps in the maintenance of patterns of methylation of DNA, while the other two are famous for initiating methylation (Shih et al. 2012). The mutation in *DNMT3A* is found in patients of CN-AML (Ostronoff et al. 2013; Roller et al. 2013; ERG 2011), and missense mutation is the most common among all, which affects R882 amino acid. These mutations are linked with high WBC count, higher age, and female gender (Roller et al. 2013). One of the epigenetic modifiers named IDH enzyme is also mutated in AML, which contains three isoforms, namely, *IDH1* (present in cytoplasm), *IDH2*, and *IDH3* (present in mitochondria), involves in the Krebs cycle, and help in the oxidative decarboxylation of isocitrate to alpha-ketoglutarate. The mutation of *IDH1* was found in 16% of CN-AML patients with *R132*, *R140*, or *R172* genes (Mardis et al. 2009) and can be utilized as an indicator for the diagnosis of genes. Additionally, the abnormalities in *IDH* have also been used in targeted therapy. For example, in their study, Wang et al. utilized *AGI-6780*, which is considered as mutant effectors of *IDH2* and induces growth in primary AML cells and TF-1 cell line. As 24% of AML patients contain aberrations in the *TET2* gene that helps in converting 5-methyl-cytosine into 5-hydroxymethyl-cytosine (Solary et al. 2014), the process of DNA demethylation is done by 5-hmc, and the mutation found in *TET2* is related to the count of white blood cells, old-age patients, and genetic mutations just like *IDH1* mutations in CN-AML. In the study by Wakita et al., aberrations found in genes that are epigenetically modified such as those in *DNMT3A*, *TET2*, and *IDH* are considered as effective biomarkers in the monitoring of MRD (Wakita et al. 2013). Most of the genes that affect posttranslational modification of histones also show mutations in AML by alterations (translocations and duplications) in the *MLL1* gene. Several genetic abnormalities in *ASXL1* are present in chronic myelomonocytic leukemia, myelodysplastic syndromes, and AML (Abdel-Wahab and Levine 2013; Tefferi 2010). Schnittger et al. in their cohort observed that mutations in *ASXL1* are mostly found in intermediary-risk AML karyotype (31%) as compared to the patients having CN-AML.

3.6.2 How Are Novel Somatic Mutations in AML Identified and Characterized?

Traditional approaches to the discovery of malignant mutations are based on sequencing of a genome, which is limited to the study of the cancer cells entering the field of cancer biology. Therefore, advanced genomic approaches come into play for the study of the entire cancer genome and find out different mutations. Initially, there is a less number of somatic variants found in AML cells, while in the protein-coding regions, there are 10–26 somatic SNVs found in four sequenced AML genomes (Ley at el. 2010; Welch et al. 2011), which concludes that AML is not the outcome of an unstable genome but might be related to numerous mutations of driver variants. The two new genes of AML are also recognized, which mutated in AML and lead to leukemogenesis. Various studies have been conducted by different researchers who identified different forms of mutations (found in *IDH1*, *IDH2*, *TET2*, *DNMT3A*, *ASXL1*, etc.) present in AML genomes, studied somewhere in published work (Welch et al. 2011; Yan et al. 2011; Schnittger et al. 2010; Yamashita et al. 2010).

3.6.3 How Are Cancer-Susceptible Genes in AML Identified?

The term "cancer susceptibility" is used for individuals having cancer at an early age, having recognized family history, or found with multiple primary cancers. More than 100 genes take part in cancer susceptibility (Garber and Offit 2005), and the process of genetic testing is not done for all these genes, because most of the genes are not identified yet. Therefore, advanced technologies of NGS have been used to detect gene mutations that lead to susceptibility of cancer. In one of the studies, for example, it has been reported that with the WGS, a woman had breast and ovarian cancers in the ages of 37 and 39, respectively, and fatal therapy-related AML when she was 42 years and had no familial history of malignancy, which suggests a syndrome of the susceptibility of cancer. The report came to negative for *BRCA1* and *BRCA2* genes when testing for gene mutations. Furthermore, WGS that was done showed that 3-kb 7–9 exons of TP53 are detected from the skin DNA of the normal patient. The results found from the genetic testing showed that cancer is not detected in the patient's mother as well as in the patient's extended family, and therefore, the study of the patient using WGS provides clinically relevant data of her relatives.

3.6.4 How Are Cryptic Translocations in Complex Genomes Characterized in AML?

For the diagnosis of AML, various cytogenetic aberrations such as translocation of t(15;17), t(8;21), and inversion of inv(16) are important and are usually used to find the response that will come with therapy. This can be studied and accomplished by using ATRA therapy in patients of AML with translocation t(15;17) and related *PML-RARA* fusion. Translocations produced by *PML-RARA* are usually not detected by only cytogenetic analysis. Various studies conducted by different researchers identified indels and translocations that involve the production of functional transcripts of the *PML-RARA* gene and have been studied somewhere in published work (Abe et al. 2008). Therefore, with our integrated NGS approach, we can easily and economically analyze clinical data of AML genomes, and it would be a better platform that opens the possibility for making treatment decisions as soon as possible based on calculated cytogenetic information and molecular features.

3.7 CHALLENGES IN THE DIAGNOSIS OF AML USING NGS

Recently developed NGS techniques have a big challenge for investigators and clinicians that face many issues and problems for the diagnosis of AML. As a large number of data are generated, people need new and advanced strategies for data curation. These methods were insufficient to detect numerous novel genetic aberrations in AML patients. Hence, most of the molecular biologists and physicians face challenges in the implementation and interpretation of molecular genetics results obtained from experiments conducted in myeloid malignancies through NGS technologies. Besides, the economic inference regarding the use of sequencing is too complex and not yet known to insurance companies, academic research, healthcare providers, and the biomedical industry. For example, Dewey et al. analyzed 56 genes from 12 fit genomes using the Illumina platform and found that out of the 56 genes, it can sequence only 51 genes with adequate coverage (Dewey et al. 2014). They concluded that there are technical challenges that need to be overcome through advanced technologies and bioinformatics. With the advancement in the field to better understand the genetic abnormalities, it is important to move toward 100% genome sequencing for diagnostics of cancer patients. We categorize challenges of using NGS technologies into two major classes, namely, biological-clinical and technical challenges.

3.7.1 Biological-Clinical Challenges

There are many challenges present in this category as follows:

- Using NGS in biological and clinical practices for distinguishing the mutated regions of *FLT3*, *CALR*, and *CEBPA* in AML patients remains challenging. Additionally, it is much difficult to utilize NGS testing times with TAT as published in the current guidelines of ELN, and hence, there is a need for conventional molecular testing for the correct diagnosis of MN.

- One of the challenges is the detection and differentiation of mutations that initiate leukemia from recurrent mutations and genetic polymorphisms such as SNPs (Stratton et al. 2009). Most of the bioinformatics techniques are unable to distinguish between driver and passenger mutations (Zhang et al. 2014).
- Another challenge is distinguishing the somatic mutations from CHIP and aberrations in pathogenic genes (Churpek et al. 2015; Pabst et al. 2008).
- Detection of reciprocal gene rearrangement at diagnosis of AML is challenging.
- Instead, growth in the implementation of NGS panels for AML, some of the TN MPN patients cannot detect any mutations.
- Inherent limitations of using NGS also persist when analyzing cancers. Approximately 500–1,000 somatic mutations are present in the AML genome, and the majority of them are not able to transform normal cells into leukemic cells. Hence, it is challenging to distinguish driver variants in the AML genome (Ley et al. 2010).

3.7.2 Technical Challenges

Although NGS is a promising and valuable method for analyzing AML as well as other cancers, the following technical limitations persist:

- The most important challenge is the data analysis which needs experienced individuals having bioinformatics skills to analyze NGS data and clinically interpret the data accurately as well as advanced computer facilities such as large data storage capabilities, specialized software, and fast data processing.
- Clinicians face the challenge of choosing between sequencing technologies, which will best suit their applications and characteristics.
- Sequence variants detected by NGS are mainly focused on the depth of reads analyzed. Most of the sequencing methods such as WGS have a coverage of approximately 30% of reads. Thus, at this limit, approximately 50% of variations are detected as false positives due to problems detected in sample preparation.
- It induces many problems in the identification and characterization of small indels by using recent innovations.
- Most of the mutations are missed due to imperfect mapping of the human genome that contains 50% of repetitive sequences.
- In AML, numbers of technical challenges are detected such as 11 deletions of mutation in *DNMT3A* have not been present on short-read analysis, and if they contain an indel, there is a problem of aligning them to the reference genome.
- The challenge for distinguishing alterations in genes by using PCR methods, that is, it is easy to diagnose correctly with a large number of transcripts, but it is somewhat difficult with low transcript numbers to distinguish genetic alterations from passenger mutations, due to problems in library preparation, sequencing process, and data analysis (Sleep et al. 2013).

- Variant calling algorithms also have an issue in the incorporation of alternate haplotypes (Schneider et al. 2017).
- NGS technologies in clinical diagnostics usually sequence the short reads of approximately 100–500 bp and have an issue of missing structural variants like longer insertions and deletions (Goodwin et al. 2016).
- Similarly, MRD identification and characterization are challenging because of the background noise of NGS.
- Due to the polyclonal nature of AML, several clones respond differently to any therapy.
- All of the NGS methods such as whole-genome sequencing are not able to give any information related to epigenetic modifications and alterations in gene expression. Hence, we have to combine the sequencing of transcriptome and bisulfate with WGS to measure the expression of RNA and identify epigenetic modifications.
- Technical challenges in clinical settings for the application of NGS also persist in incorrect identification of repetitive sequences and capture of GC (guanine-cytosine)-imbalanced targets, and the time needed for interpreting the data will require a skilled bioinformatician to analyze and clinically interpret NGS data accurately.

3.7.3 Other Challenges

As of now, the sequencing of the genome remains costly, and it is challenging for those who use NGS regularly in clinical settings due to the unavailability of expertise, infrastructure, and time to complete analysis. With the rise in the necessity of sequencing techniques day by day, there is a special need for technical and bioinformatics infrastructure for the analysis and validation of clinical data. The high financial burden of the use of NGS is also a great challenge as most of the expenses are on the technical and staff equipment such as genetic counselors, specialized clinicians, and molecular and computational biologists. Although the costs of NGS are decreasing day by day, there is a challenge for individuals of time for processing the huge datasets of sequencing for obtaining diagnostic results and taking treatment decisions.

3.8 FUTURE OF NGS IN CANCER GENOMES

As we know, the recent high-throughput NGS methods have detected numerous genetic abnormalities in cancer genomes. Recently, the sequencing techniques help in the identification and characterization of most of the AML mutations, but there exist many challenges that have to be overcome soon. It was identified that NGS techniques relevant to gene panel are the most suitable methods for molecular characterization of AML patients, but still there remain many drawbacks. Hence, it will need to be validated individually in all diagnostic laboratories. The technical challenges in clinical settings for the application of NGS include incorrect sequencing of repetitive sequences and capture of GC-imbalanced targets; the time needed for interpreting the data will require a skilled bioinformatician to analyze and clinically interpret the NGS data accurately; and advanced computer facilities such as large data storage capabilities, specialized software, and fast data processing are needed. Therefore, the collaboration is made among bioinformaticians, scientists, and physicians that will create a clinical database, which helps in the clinical settings in the laboratories. It is useful to know the evolution of AML genome, as it will help in the interpretation of relapse mechanisms and therapeutic resistance. Molecular MRD diagnostics should be used in near future for identifying patient response to therapy. Most of the gene mutations in AML suggest renewing the ELN classification of AML in the future. There should be a single platform of NGS panel for any cancer genome that will simultaneously analyze and validate indels, SNVs, aberrant gene expression, CNVs, and common gene rearrangements. When strand bias per amplicon and mean read depth are being assessed, most of the sequencing fragments show good-quality reads, but still some of the regions that contain high CG contents show low performance and cannot be assessed by NGS technologies and conventional molecular methods. However, these limitations can be solved by using the Integrative Genomics Viewer (IGV). There is a large gap of knowledge that exists in the assay of CNV karyotyping, and hence, this has to be removed by validating the sensitivity of CNV karyotyping assay for identification of gain and loss of any region of the chromosome. We can classify the AML by using NGS techniques. The molecular classification of AML is the best option to identify mutations in genes, and therefore the molecular classification will be further used to access genetic mutations in the AML genome. We can optimize the therapy of AML by knowing all the mutations present in the genome of AML patients. Hence, due to the characterization of AML mutations, we will develop a new way to identify therapeutic targets. The absorption, distribution, metabolism, excretion, and toxicity (ADMET) properties and efficiency of any drug are influenced by large variations in genes, which influence the toxicity and efficacy of a therapeutic substance and give the correct knowledge regarding the dose and choice of chemotherapy used. Hence, due to the enhancement of sequenced genomes of AML patients, the genetic variants affecting chemotherapy will also be increased. NGS may identify genetic variants of more than 100 genes in AML patients and contribute to cancer susceptibility. But the causal associations among different variants and AML are not likely to show for all genes, and hence, the variations in genes contributing to susceptibility in AML patients will be enhanced in near future. Recently, it has been noted that most of the novel drugs targeted to epigenetic changes have been in clinical trials in cell lines of leukemia and also in preclinical leukemia outcomes, which provide the potential therapeutics against mutations in AML leading to possibly better treatment outcomes. Additionally, the incorporation of deep sequencing methods of NGS has opened a new way for personalized medicine. We need in vitro and in vivo experimental models for exhaustive functional studies, which will further help in better understanding the biological importance of genetic mutations and hence will help in distinguishing between drivers and mutations. Even after the assessment and validation of mutant genes, a huge number of false positives are left behind, and hence, the analysis and validation of those genes require high levels of sensitivity for a specific sample. Molecular barcode-based NGS panels should be used, which can detect a large number of mutations and provide the best option for the treatment of disease. Targeted therapeutics will also be developed for the specific identification of disease. In a study by Dewey et al. (2014), in 56 genes analyzed from 12 fit genomes using the Illumina platform, only 51 genes with adequate coverage could be sequenced, and hence, there is a technical challenge which needs to be overcome through advanced technologies and bioinformatics. However, soon the sequencing of AML mutations could be performed economically with the help of advanced NGS technologies (third-generation technologies). With the advancement in the field to better understand the genetic abnormalities, it is important to move toward 100% genome sequencing for diagnostics of cancer patients.

3.9 CONCLUSION

The accomplishment of the HGP and the discovery of recent NGS techniques leads to a revolution in genomic research that allows rapid and accurate sequencing of most of the genes of any genome at low cost within a day. These NGS technologies pave the way for several applications and can produce numerous repeats in a single run and hence display numerous novel genetic alterations to understand and characterize cancers at a molecular level. This has brought to light the perception of the complex genome of AML in identifying mutations in the genome such as *MLLT3*, *DNTM3A*, *RUNX1*, *KMT2A*, *CBFB*, *PML-RARA*, *DEK*, *MECOM*, *GATA2*, *CEPBA* (biallelic), *NPM1*, *ASXL1*, *TP53*, *IDH*, *FLT3*, *TET2*, *SRSF2*, *TP53*, *c-KIT*, and *SF3B1*, risk stratification, classification, diagnosis, and assessing response to AML, which have been subsequently utilized in prognostic subclassification of AML. Additionally, these techniques are utilized for understanding different areas of genome and transcriptome such as DNA methylation, polymorphism detection, histone modifications, mutation mapping, protein–protein interactions, DNA–protein interactions, personal genomics, and treatment of common disorders at molecular level. The AML patients aged more than 60 years have a lower incidence of favorable cytogenetics and are loaded with a high number of molecular mutations as compared to their younger counterparts. There may persist biological-clinical challenges as well as the technical difficulties in integrating the knowledge of the application of NGS into clinical settings of AML, which can be resolved by applying conventional molecular testing for the correct diagnosis of malignancies, recruiting skilled bioinformaticians to analyze NGS data, clinically interpreting the data accurately, and having advanced computer facilities for data storage and processing. The NGS-based AML sequencing has paved the way for looking into the biological pathogenesis of leukemia, which provides the potential therapeutics against mutations in AML leading to possibly better treatment outcomes. Additionally, the incorporation of deep sequencing methods of NGS has opened a new way for personalized medicine and provides information on tumor pathogenesis, useful novel biomarkers, and diagnostics and targeted therapeutics of myeloid malignancies. However, with continuous improvements and use of NGS, it is considered as a regular method for the treatment of AML and other malignancies.

REFERENCES

Abdel-ahab, O., & Levine, R. L. (2013). Mutations in epigenetic modifiers in the pathogenesis and therapy of acute myeloid leukaemia. *Blood: The Journal of the American Society of Hematology*, 121(18), 3563–3572.

Abe, S., Ishikawa, I., Harigae, H., & Sugawara, T. (2008). A new complex translocation t (5; 17; 15)(q11; q21; q22) in acute promyelocytic leukaemia. *Cancer Genetics and Cytogenetic*, 184(1), 44–47.

Appelbaum, F. R., Gundacker, H., Head, D. R., Slovak, M. L., Willman, C. L., Godwin, J. E., … & Petersdorf, S. H. (2006). Age and acute myeloid leukaemia. *Blood*, 107(9), 3481–3485.

Biesecker, L. G., Burke, W., Kohane, I., Plon, S. E., & Zimmern, R. (2012). Next-generation sequencing in the clinic: are we ready? *Nature Reviews Genetics*, 13(11), 818–824.

Burnett, A., Wetzler, M., & Lowenberg, B. (2011). Therapeutic advances in acute myeloid leukaemia. *Journal of Clinical Oncology*, 29(5), 487–494.

Burrell, R. A., & Swanton, C. (2014). The evolution of the unstable cancer genome. *Current Opinion in Genetics & Development*, 24, 61–67.

Campos, L., Guyotat, D., Archimbaud, E., Devaux, Y., Treille, D., Larese, A., … & Fiere, D. (1989). Surface marker expression in adult acute myeloid leukaemia: correlations with initial characteristics, morphology and response to therapy. *British Journal of Haematology*, 72(2), 161–166.

Castaigne, S., Pautas, C., Terré, C., Raffoux, E., Bordessoule, D., Bastie, J. N., … & de Revel, T. (2012). Effect of gemtuzumab ozogamicin on survival of adult patients with de-novo acute myeloid leukaemia (ALFA-0701): a randomised, open-label, phase 3 study. *The Lancet*, 379(9825), 1508–1516.

Cheson, B. D., Bennett, J. M., Kopecky, K. J., Büchner, T., Willman, C. L., Estey, E. H., … & Lo-Coco, F. (2003). Revised recommendations of the international working group for diagnosis, standardization of response criteria, treatment outcomes, and reporting standards for therapeutic trials in acute myeloid leukaemia. *Journal of Clinical Oncology*, 21(24), 4642–4649.

Churpek, J. E., Pyrtel, K., Kanchi, K. L., Shao, J., Koboldt, D., Miller, C. A., … & Pusic, I. (2015). Genomic analysis of germ line and somatic variants in familial myelodysplasia/acute myeloid leukaemia. *Blood: The Journal of the American Society of Hematology*, 126(22), 2484–2490.

DeVita J, V. T., & Rosenberg, S. A. (2012). Two hundred years of cancer research. *New England Journal of Medicine*, 366(23), 2207–2214.

Dewey, F. E., Grove, M. E., Pan, C., Goldstein, B. A., Bernstein, J. A., Chaib, H., … & Pakdaman, N. (2014). Clinical interpretation and implications of whole-genome sequencing. *JAMA*, 311(10), 1035–1045.

Ding, L., Ley, T. J., Larson, D. E., Miller, C. A., Koboldt, D. C., Welch, J. S., … & McMichael, J. F. (2012). Clonal evolution in relapsed acute myeloid leukaemia revealed by whole-genome sequencing. *Nature*, 481(7382), 506–510.

Döhner, H., & Gaidzik, V. I. (2011). *Impact of Genetic Features on Treatment Decisions in AML. Hematology 2010*, the American Society of Hematology Education Program Book, 2011(1), 36–42.

Döhner, H., Estey, E., Grimwade, D., Amadori, S., Appelbaum, F. R., Büchner, T., … & Levine, R. L. (2017). Diagnosis and management of AML in adults: 2017 ELN recommendations from an international expert panel. *Blood*, 129(4), 424–447.

Dombret, H., Preudhomme, C., & Boissel, N. (2009). Core binding factor acute myeloid leukaemia (CBF-AML): is high-dose Ara-C (HDAC) consolidation as effective as you think? *Current Opinion in Hematology*, 16(2), 92–97.

Dovichi, N. J. (1997). DNA sequencing by capillary electrophoresis. *Electrophoresis*, 18(12–13), 2393–2399.

Duncavage, E. J., Abel, H. J., Szankasi, P., Kelley, T. W., & Pfeifer, J. D. (2012). Targeted next generation sequencing of clinically significant gene mutations and translocations in leukaemia. *Modern Pathology*, 25(6), 795–804.

ERG, E. (2011). Incidence and prognostic influence of DNMT3A mutations in acute myeloid leukaemia. *Journal of Clinical Oncology*, 29, 2889–2896.

Estey, E., & Döhner, H. (2006). Acute myeloid leukaemia. *The Lancet*, 368(9550), 1894–1907.

Fukuda, M., Asano, S., Nakamura, T., Adachi, M., Yoshida, M., Yanagida, M., & Nishida, E. (1997). CRM1 is responsible for intracellular transport mediated by the nuclear export signal. *Nature*, 390(6657), 308–311.

Gaidzik, V., & Döhner, K. (2008, August). Prognostic implications of gene mutations in acute myeloid leukaemia with normal cytogenetic. Seminars in Oncology, 35(4), 346–355.

Garber, J. E., & Offit, K. (2005). Hereditary cancer predisposition syndromes. *Journal of Clinical Oncology*, 23(2), 276–292.

Gerstung, M., Papaemmanuil, E., Martincorena, I., Bullinger, L., Gaidzik, V. I., Paschka, P., … & Ganser, A. (2017). Precision oncology for acute myeloid leukaemia using a knowledge bank approach. *Nature Genetics*, 49(3), 332.

Goodwin, S., McPherson, J. D., & McCombie, W. R. (2016). Coming of age: ten years of next-generation sequencing technologies. *Nature Reviews Genetics*, 17(6), 333.

Grimwade, D., & Freeman, S. D. (2014). Defining minimal residual disease in acute myeloid leukaemia: which platforms are ready for "prime time"? *Blood: The Journal of the American Society of Hematology*, 124(23), 3345–3355.

Grossmann, V., Tiacci, E., Holmes, A. B., Kohlmann, A., Martelli, M. P., Kern, W., … & Trifonov, V. (2011). Whole-exome sequencing identifies somatic mutations of BCOR in acute myeloid leukaemia with normal karyotype. *Blood: The Journal of the American Society of Hematology*, 118(23), 6153–6163.

Hansen, E., Quivoron, C., Straley, K., Lemieux, R. M., Popovici-Muller, J., Sadrzadeh, H., … & Micol, J. B. (2014). AG-120, an oral, selective, first-in-class, potent inhibitor of mutant IDH1, reduces intracellular 2HG and induces cellular differentiation in TF-1 R132H cells and primary human IDH1 mutant AML patient samples treated ex vivo.

Hasserjian, R. P. (2013). Acute myeloid leukaemia: advances in diagnosis and classification. *International Journal of Laboratory Hematology*, 35(3), 358–366.

Jabeen, A., Ahmad, N., & Raza, K. (2018). Machine learning-based state-of-the-art methods for the classification of rna-seq data. In *Classification in BioApps* (pp. 133–172). Springer, Cham.

Kihara, R., Nagata, Y., Kiyoi, H., Kato, T., Yamamoto, E., Suzuki, K., … & Miyazaki, Y. (2014). Comprehensive analysis of genetic alterations and their prognostic impacts in adult acute myeloid leukaemia patients. *Leukaemia*, 28(8), 1586–1595.

Klepin, H. D. (2014). *Geriatric Perspective: How to Assess Fitness for Chemotherapy in Acute Myeloid Leukaemia. Hematology 2014*, the American Society of Hematology Education Program Book, 2014(1), 8–13.

Koboldt, D. C., Steinberg, K. M., Larson, D. E., Wilson, R. K., & Mardis, E. R. (2013). The next-generation sequencing revolution and its impact on genomics. *Cell*, 155(1), 27–38.

Ley, T. J., Ding, L., Walter, M. J., McLellan, M. D., Lamprecht, T., Larson, D. E., … & Harris, C. C. (2010). DNMT3A mutations in acute myeloid leukaemia. *New England Journal of Medicine*, 363(25), 2424–2433.

Ley, T. J., Mardis, E. R., Ding, L., Fulton, B., McLellan, M. D., Chen, K., … & Cook, L. (2008). DNA sequencing of a cytogenetically normal acute myeloid leukaemia genome. *Nature*, 456(7218), 66–72.

Lindsley, R. C., Mar, B. G., Mazzola, E., Grauman, P. V., Shareef, S., Allen, S. L., … & Damon, L. E. (2015). Acute myeloid leukaemia ontogeny is defined by distinct somatic mutations. *Blood, The Journal of the American Society of Hematology*, 125(9), 1367–1376.

Lister, R., Pelizzola, M., Dowen, R. H., Hawkins, R. D., Hon, G., Tonti-Filippini, J., … & Edsall, L. (2009). Human DNA methylomes at base resolution show widespread epigenomic differences. *Nature*, 462(7271), 315–322.

Marcucci, G., Haferlach, T., & Döhner, H. (2011). Molecular genetics of adult acute myeloid leukaemia: prognostic and therapeutic implications. *Journal of Clinical Oncology*, 29(5), 475–486.

Mardis, E. R., Ding, L., Dooling, D. J., Larson, D. E., McLellan, M. D., Chen, K., … & Fulton, L. A. (2009). Recurring mutations found by sequencing an acute myeloid leukaemia genome. *New England Journal of Medicine*, 361(11), 1058–1066.

McGettigan, P. A. (2013). Transcriptomics in the RNA-seq era. *Current Opinion in Chemical Biology*, 17(1), 4–11.

Morozova, O., & Marra, M. A. (2008). Applications of next-generation sequencing technologies in functional genomics. *Genomics*, 92(5), 255–264.

Ostronoff, F., Othus, M., Ho, P. A., Kutny, M., Geraghty, D. E., Petersdorf, S. H., … & Stirewalt, D. L. (2013). Mutations in the DNMT3A exon 23 independently predict poor outcome in older patients with acute myeloid leukaemia: a SWOG report. *Leukaemia*, 27(1), 238–241.

Pabst, T., Eyholzer, M., Haefliger, S., Schardt, J., & Mueller, B. U. (2008). Somatic CEBPA mutations are a frequent second event in families with germline CEBPA mutations and familial acute myeloid leukaemia. *Journal of Clinical Oncology*, 26(31), 5088–5093.

Papaemmanuil, E., Gerstung, M., Bullinger, L., Gaidzik, V. I., Paschka, P., Roberts, N. D., … & Gundem, G. (2016). Genomic classification and prognosis in acute myeloid leukaemia. *New England Journal of Medicine*, 374(23), 2209–2221.

Patel, J. P., Gönen, M., Figueroa, M. E., Fernandez, H., Sun, Z., Racevskis, J., … & Huberman, K. (2012). Prognostic relevance of integrated genetic profiling in acute myeloid leukaemia. *New England Journal of Medicine*, 366(12), 1079–1089.

Quintás-ardama, A., Ravandi, F., Liu-Dumlao, T., Brandt, M., Faderl, S., Pierce, S., … & Kantarjian, H. (2012). Epigenetic therapy is associated with similar survival compared with intensive chemotherapy in older patients with newly diagnosed acute myeloid leukaemia. *Blood: The Journal of the American Society of Hematology*, 120(24), 4840–4845.

Raza, K., & Ahmad, S. (2019). Recent advancement in next-generation sequencing techniques and its computational analysis. *International Journal of Bioinformatics Research and Applications*, 15(3), 191–220.

Roller, A., Grossmann, V., Bacher, U., Poetzinger, F., Weissmann, S., Nadarajah, N., … & Haferlach, T. (2013). Landmark analysis of DNMT3A mutations in hematological malignancies. *Leukaemia*, 27(7), 1573–1578.

Rücker, F. G., Schlenk, R. F., Bullinger, L., Kayser, S., Teleanu, V., Kett, H., … & Paschka, P. (2012). TP53 alterations in acute myeloid leukaemia with complex karyotype correlate with specific copy number alterations, monosomal karyotype, and dismal outcome. *Blood: The Journal of the American Society of Hematology*, 119(9), 2114–2121.

Sanger, F., Nicklen, S., & Coulson, A. R. (1977). DNA sequencing with chain-terminating inhibitors. *Proceedings of the National Academy of Sciences*, 74(12), 5463–5467.

Schiffer, C. A. (2014). Optimal dose and schedule of consolidation in AML: is there a standard? *Best Practice & Research Clinical Haematology*, 27(3–4), 259–264.

Schneider, V. A., Graves-Lindsay, T., Howe, K., Bouk, N., Chen, H. C., Kitts, P. A., … & Fulton, R. S. (2017). Evaluation of GRCh38 and de novo haploid genome assemblies demonstrates the enduring quality of the reference assembly. *Genome Research*, 27(5), 849–864.

Schnittger, S., Haferlach, C., Ulke, M., Alpermann, T., Kern, W., & Haferlach, T. (2010). IDH1 mutations are detected in 6.6% of 1414 AML patients and are associated with intermediate risk karyotype and unfavorable prognosis in adults younger than 60 years and unmutated NPM1 status. *Blood*, 116(25), 5486–5496.

Shih, A. H., Abdel-Wahab, O., Patel, J. P., & Levine, R. L. (2012). The role of mutations in epigenetic regulators in myeloid malignancies. *Nature Reviews Cancer*, 12(9), 599–612.

Sleep, J. A., Schreiber, A. W., & Baumann, U. (2013). Sequencing error correction without a reference genome. *BMC Bioinformatics*, 14(1), 367.

Solary, E., Bernard, O. A., Tefferi, A., Fuks, F., & Vainchenker, W. (2014). The Ten-Eleven Translocation-2 (TET2) gene in hematopoiesis and hematopoietic diseases. *Leukaemia*, 28(3), 485–496.

Sorror, M. L., Maris, M. B., Storb, R., Baron, F., Sandmaier, B. M., Maloney, D. G., & Storer, B. (2005). Hematopoietic cell transplantation (HCT)-specific comorbidity index: a new tool for risk assessment before allogeneic HCT. *Blood,* 106(8), 2912–2919.

Stratton, M. R., Campbell, P. J., & Futreal, P. A. (2009). The cancer genome. *Nature*, 458(7239), 719–724.

Sudhindra, A., & Smith, C. C. (2014). FLT3 inhibitors in AML: are we there yet? *Current Hematologic Malignancy Reports*, 9(2), 174–185.

Tefferi, A. (2010). Novel mutations and their functional and clinical relevance in myeloproliferative neoplasms: JAK2, MPL, TET2, ASXL1, CBL, IDH and IKZF1. *Leukaemia*, 24(6), 1128.

Vardiman, J. W., Thiele, J., Arber, D. A., Brunning, R. D., Borowitz, M. J., Porwit, A., … & Bloomfield, C. D. (2009). The 2008 revision of the World Health Organization (WHO) classification of myeloid neoplasms and acute leukaemia: rationale and important changes. *Blood: The Journal of the American Society of Hematology*, 114(5), 937–951.

Wakita, S., Yamaguchi, H., Omori, I., Terada, K., Ueda, T., Manabe, E., … & Todoroki, T. (2013). Mutations of the epigenetics-modifying gene (DNMT3a, TET2, IDH1/2) at diagnosis may induce FLT3-ITD at relapse in de novo acute myeloid leukaemia. *Leukaemia*, 27(5), 1044–1052.

Welch, J. S., Westervelt, P., Ding, L., Larson, D. E., Klco, J. M., Kulkarni, S., … & Veizer, J. (2011). Use of whole-genome sequencing to diagnose a cryptic fusion oncogene. *JAMA*, 305(15), 1577–1584.

Wen, H., Li, Y., Malek, S. N., Kim, Y. C., Xu, J., Chen, P., … & Mankala, S. (2012). New fusion transcripts identified in normal karyotype acute myeloid leukaemia. *PloS One*, 7(12).

Yamashita, Y., Yuan, J., Suetake, I., Suzuki, H., Ishikawa, Y., Choi, Y. L., … & Takada, S. (2010). Array-based genomic resequencing of human leukaemia. *Oncogene*, 29(25), 3723–3731.

Yan, X. J., Xu, J., Gu, Z. H., Pan, C. M., Lu, G., Shen, Y., … & Liang, W. X. (2011). Exome sequencing identifies somatic mutations of DNA methyltransferase gene DNMT3A in acute leukaemia. *Nature Genetics*, 43(4), 309.

Zhang, J., Liu, J., Sun, J., Chen, C., Foltz, G., & Lin, B. (2014). Identifying driver mutations from sequencing data of heterogeneous tumours in the era of personalized genome sequencing. *Briefings in Bioinformatics*, 15(2), 244–255.

4 Translational Bioinformatics Methods for Drug Repurposing

Jyoti Rani and Urmi Bajpai
University of Delhi

Srinivasan Ramachandran
Council of Scientific and Industrial Research – Institute of Genomics and Integrative Biology (CSIR-IGIB)

CONTENTS

4.1 INTRODUCTION

Our healthcare system is facing challenges such as decreasing number of new chemical entities in the discovery and development of new therapeutics accompanied by rising costs (Readhead and Dudley 2013). Many pathogens are becoming resistant to the existing drugs (Khatoon, Alam et al. 2019) and urge the need for new drugs. The overall process of new drug development takes more than 12 years (DiMasi, Feldman et al. 2010) and the estimated cost reaches 2.6 billion dollars (DiMasi, Grabowski et al. 2016). The scientific discovery of a biological target for a given disease is the first step in the development of a novel drug (Mohs and Greig 2017). Sometimes it is necessary to produce multiple drug candidates, as various molecules fail in different stages due to lack of safetyor potencyor kinetics, a combination of multiple factors (Mohs and Greig 2017). The overall process of discovery and development of a drug is depicted in Figure 4.1. The clinical phase is divided into four sub phases, namely, phase I, phase II, phase III, and phase IV. In Phase I, human safety is evaluated, in phase II, the efficacy and safety are evaluated. Phase III focuses on safety and toxicity in a large population. Phase IV, also known as the pharmacovigilance phase, evaluates the parameters such as its side effects, interactions with drugs, or any adverse effect of the drug.

4.1.1 Drug Repurposing

These above-discussed challenges can be addressed by finding new indications of already approved drugs; this process is called "drugrepurposing" (DR) (Pushpakom, Iorio et al. 2019). In the literature, various terminologies are used for DR, namely, "drug repositioning," "drug reformulation," and "drug reprofiling" (Langedijk, Mantel-Teeuwisse et al. 2015). Drug repositioning tends to find the new therapeutic indications of approved drug candidates with different pharmacological targets. Drug reformulation is the process of developing new formulations of drugs by using technological advances (Murteira, Ghezaiel et al. 2013). Drug combination refers to combining one or more drugs formerly used as solo pharmaceuticals.

The DR is considered more advantageous over developing new drugs in terms of time frame, low risk of failure, and less budget requirement (Pushpakom, Iorio et al. 2019). The approved drugs for repurposing are safe, as all the early-stage trials have been done and the chance to fail in further efficacy trials would be less. The

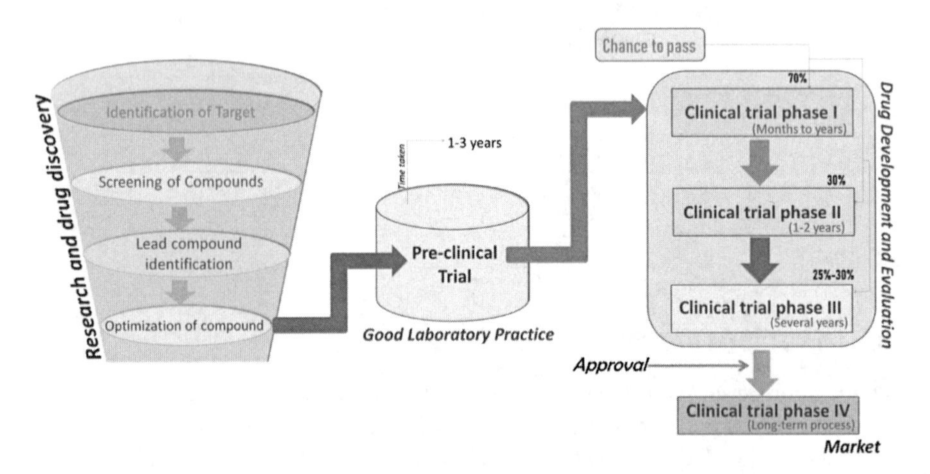

FIGURE 4.1 The schematic representation of the progressions in the discovery and development of a drug.

total time required for the development of a new drug can be reduced as the drug has already undergone preclinical trials, safety assessments, and formulation development. Most importantly, cost reduction is an advantage, which depends on the stage of the development (Breckenridge and Jacob 2019). The comparison between the novel drug development process and the DR process is depicted in Figure 4.2. The repurposing of a drug candidate may also lead to the identification of new targets and pathways for further investigation (Pushpakom, Iorio et al. 2019).

Originally, this methodology was used for serendipitous drug discovery and the prominent examples are Sildenafil and Thalidomide drugs; however, their repurposing does not follow any systematic approach (Ashburn and Thor 2004). The drug Sildenafil was designed in 1980 while searching a drug for angina but approved for the treatment of erectile dysfunction (Boolell, Allen et al. 1996). Sildenafil effectively inhibits the phosphodiesterase type 5, enhances the cGMP (cyclic guanosine3′,5′ monophosphate) concentration, and leads to the amplification of the endogenous nitric oxide-cGMP signaling pathway (Hatzimouratidis 2006).It is well tolerated, safe to use, and gives satisfactory results to the patients. Thalidomide was marketed in 1957 for the treatment of morning sickness in pregnant women, but its consumption caused severe skeletal birth defects. Lately, in 1964, the same drug was accidentally revived by Jacob Sheskin to treat erythema nodosumlaprosum (Anderson 2001). Many years later, thalidomide was reported as an inhibitor of tumor necrosis factor α (Sampaio, Sarno et al. 1991). In 1993, Gilla Kaplan discovered that it could suppress the activation of latent HIV (human immunodeficiency virus) [PMC46849]. The repurposing of thalidomide also led to the discovery of its very effective derivatives, for example, lenalidomide (Urquhart 2018), with antineoplastic, antiangiogenic, pro-erythropoietic, and immune-modulatory properties (Chen, Zhou et al. 2017). Few examples of DR are given in Table 4.1.

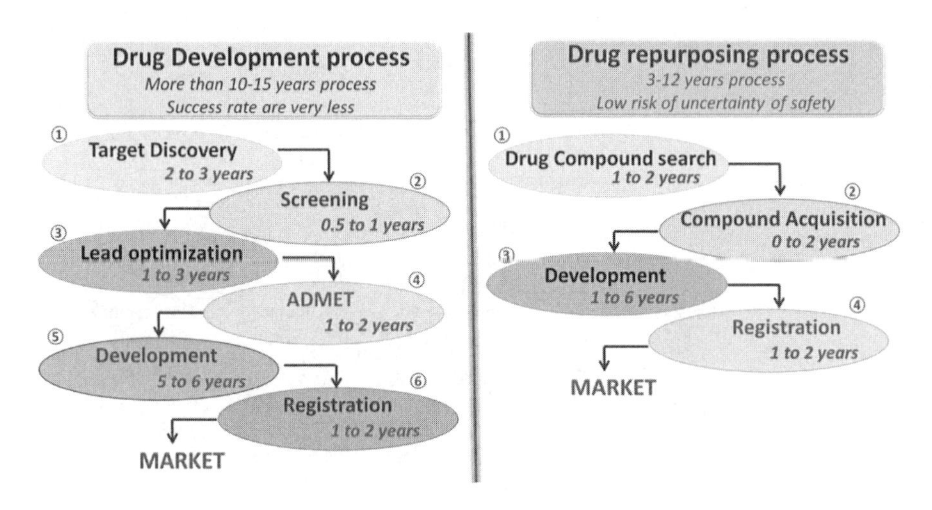

FIGURE 4.2 Comparison between drug development and drug repurposing processes. (Adapted from Ashburn and Thor 2004.)

TABLE 4.1
Examples of Repurposed Drugs

Drugs	Original Indication	Original MoA	Repurposed Indication	Modified MoA
Amantadine	Influenza A virus	Antiviral activity	Parkinson's disease	Releases dopamine from the nerve endings of the brain cells and also stimulate the norepinephrine response
Amphotericin B	Paralysis	Reverser of neuromuscular blockade	Alzheimer's disease	By binding to the neuronal nicotinic acetylcholine receptors (nAChR) at a binding site that is different from acetylcholine binding site
Mecamylamine	Hypertension	Nicotinic parasympathetic ganglionic blocker	Depression	Neuronal nicotinic receptor modulator with antidepressant activity
Itraconazole	Fungal infection	Inhibits the enzyme cytochrome P450 14α-demethylase	Anticancer	Inhibits endothelial cell cholesterol trafficking and angiogenesis
Nelfinavir	Human immunodeficiency virus	HIV-1 protease inhibitor	Anticancer	Inhibits 20S proteosome. AKT signaling, HSP90, and HER2 signaling
Valsartan	Hypertension	Binds to angiotensin receptor 1 and inhibit the binding of angiotensin II	Alzheimer's disease	Lowers brain beta-amyloid protein levels and improves spatial learning
Tamoxifen	Breast cancer	Inhibits estrogen binding to its receptor	Bipolar disorder	The protein kinase C activity is inhibited
Methotrexate	Cancer	Inhibits the enzymes responsible for nucleotide synthesis	Psoriasis	Interferes with the inflammatory pathways critical to psoriasis pathogenesis

DR also brings some challenges along with the abovementioned advantages. The preclinical and clinical trials, however, cannot be skipped due to dissatisfactory drug information or obsolete drug entries (Ashburn and Thor 2004). Another common complication is less efficacy or high toxicity of drugs in new therapeutic indications. Sometimes the identified and tolerated concentrations of the drugs are either lower or higher than the requirement. Two or more drugs with diverse mechanisms of actions (MoA) can be combined as an alternative to achieve success in DR (Sun, Sanderson et al. 2016).

4.1.1.1 Drug Repurposing Profiles

The ON-TARGET and OFF-TARGET are two widely known profiles for DR (de Oliveira and Lang 2018). In an ON-TARGET profile, the new therapeutic indication can be found by considering the known pharmacological mechanism of the drug compound, meaning that molecules acting on the same biological target treat different diseases (Koch, Hamacher et al. 2014). For example, the drug finasteride for male-pattern baldness (alopecia) (de Oliveira and Lang 2018) was designed and developed for the treatment of "benign prostatic hyperplasia" by reducing the bioavailability of dihydrotestosterone, an active metabolite of testosterone. Testosterone is accountable for the growth of the prostate by inhibiting the 5-α-reductase enzyme. The reduced bioavailability of dihydrotestosterone is also noticed in the scalp by using finasteride, which further leads to inhibiting the miniaturization of the hair follicle and aids in the treatment of alopecia (de Oliveira and Lang 2018). The drug finasteride acts on the same target but yields two different therapeutic effects (Ashburn and Thor2004; Mohs and Greig2017). *Thalidomide* is an example of the OFF-TARGET profile, wherein the drug acts on a new target for new indications.

In DR, the drug–disease relationship can be detected using a variety of methodologies, such as "drugoriented," "targetoriented," and "diseaseoriented" (Jin and Wong 2014). The application or selection of a given methodology further depends on the available biological information, that is, pharmacological and toxicological information. In "drug-oriented" methodology, the drug compound (molecule) is screened for its structural characteristics, adverse effects, and phenotypic properties and leads to the identification of the ability to cause desirable changes in the specific phenotype (Moffat, Vincent et al. 2017). This strategy can also be termed as "blind research approach," where new indications of an old drug are observed at random (Ashburn and Thor 2004). The "target-oriented" approach encompasses in vitro and in vivo high-throughput screening of drugs for a target protein and also includes the in silico screening of drug libraries (Jin and Wong 2014; Bellomo, Medina et al. 2017). Another approach is "diseaseoriented," wherein detailed information about modulation of the disease phenotype by the drugs are used. These details include genomics, proteomics, and metabolomics information. This approach is generally accessed by using computational methodologies such as interacting networks and pathway modeling (Li and Jones 2012; Jadamba and Shin 2016; Wu, Huang et al. 2017). Using this methodology offers an understanding of omics data and leads to the discovery of disease pathways and possible pharmacological targets.

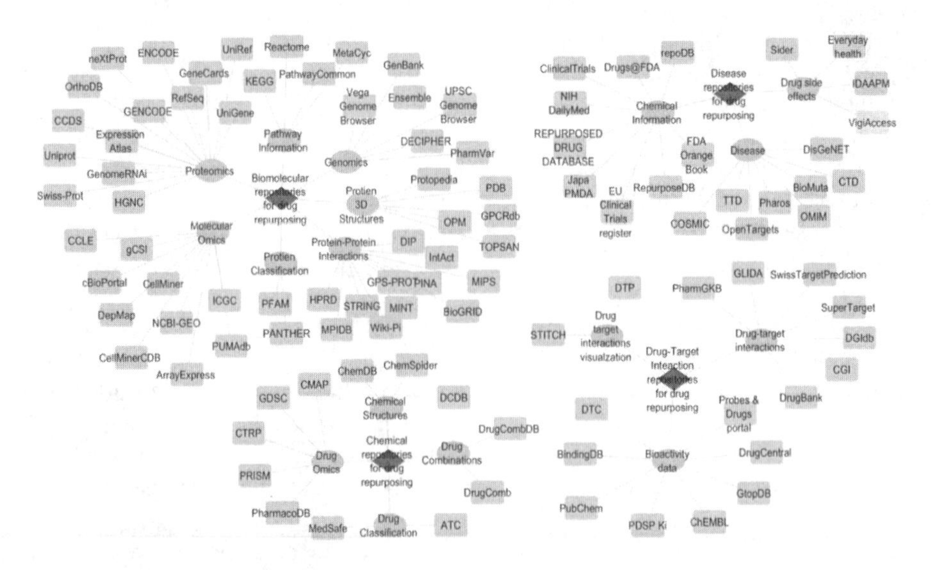

FIGURE 4.3 Distribution of publicly available repositories for drug repurposing. The diamonds represent the four major categories of databases; circles are the subcategories of the databases. The rectangles are the databases under these categories and subcategories. (Data adapted from Tanoli, Seemab et al. 2020. The tabular representation of data is shown in Table 4.2.)

4.1.2 Databases/Repositories for Drug Repurposing

In the literature, several reviews have reported various tools and repositories for DR. In 2011, Dudley and Deshpande discussed computational methodologies for DR (Dudley, Deshpande et al. 2011). Jin and Wong, in 2014, published the methodologies to link the existing DR methods with biological and pharmaceutical information to further advance the pipeline (Jin and Wong 2014). Likewise, various Web-based tools (Sam and Athri 2019), resources, and also the application of artificial intelligence (AI) in DR have been published (Yang, Wang et al. 2019). The databases with free access are very useful in the repurposing methodologies. Several chemicals, biomolecular, drug–target interaction (DTI), and disease databases are publicly available for DR (Tanoli, Seemab et al. 2020). The available repositories for DR are shown in Figure 4.3. The publicly available repositories for DR are classified into 4 categories, which are further subdivided into 17 subcategories.

The rising rate of creation of these databases is accelerating the drug development in new ways (Hodos, Kidd et al. 2016). The available information from these repositories further serves in the development of computational approaches for DR (Xue, Li et al. 2018). In comparison with biotechnological experiments, bioinformatics approaches could lower costs and face fewer obstacles in the search for new indications of already approved drugs (Oprea and Overington 2015). Computational approaches are developed for DR (de Oliveira and Lang 2018) and offer the exhaustive search of all potential candidates. Most computational efforts are driven by large experimental data stored in public repositories.

TABLE 4.2
The Publicly Available Repositories for Drug Repurposing

Database Categories	Subcategories	Databases
Chemical repositories for drug repurposing	Drug combinations Drug classification Chemical structures Drug omics	DrugComb, DrugCombDB, DCDB, MedSafe, ATC, ChemSpider, ChemDB, CMAP, CTRP, PRISM, PharmacoDB
Biomolecular repositories for drug repurposing	Genomics Proteomics Protein–protein interactions Molecular omics Pathway information Protein 3D structures Protein classification	Vega Genome Browser, UCSC Genome Browser, Ensemble, GenBank, DECIPHER, PharmVar, UniProt, GENCODE, ENCODE, CCDS, GenCards, GenomeRNAi, HGNC, OrthoDB, UniGene, UnieRef, NextProt, RefSeq, Expression Atlas, HPRD, BioGRID, MINT, GPS-PROT, Wiki-Pi, PINA, MPIDB, STRING, MIPS, IntAct, DIP, CCLE, gCCSI, DepMap, INGC, cBioPortal, NCBI-GEO, ArrayExpress, PUMAdb, CellMiner, CellMinerCDB, PathwayCommon, KEGG, Reactome, MetaCyc, PDB, OPM, Protopedia, TOPSAN, GPCRdb, PFAM, PANTHER
Drug–target interaction repositories for drug repurposing	Bioactivity data Drug–target interactions Drug–target interactions visualization	DTP, STITCH, DrugBank, PharmGKB, SupetTarget, GLIDA, SwissTargetPrediction, CGI, DTC, ChEMBL, PubChem, BinidngDB, DrugCentral, GTopDB, PDSP Ki, Probes & Drugs Portal
Disease repositories for drug repurposing	Disease Chemical information Drug side effects	Sider, IDAAPM, Everyday health, VigiAccess, ClinicalTrials, Drugs@FDA, FDA Orange Book, EU Clinical Trilasregister, Japa PMDA, NIH DailyMed, Repurposed Drug Database, repoDB, RepurposeDB, OpenTargets, COSMIC, Pharos, DisGeNET, BioMuta, TTD, OMIM, CTD

4.1.2.1 Usage of Above Repositories in Drug Repurposing

The multidimensional data available in the repositories can be used for DR. An example framework of one possible methodology for generating a hypothesis for DR for disease "A" is displayed in Figure 4.4. Let us start with the clinical databases for the knowledge of approved drugs for disease "A." The clinical databases are ClinicalTrials (https://clinicaltrials.gov/), OrangeBook (https://www.accessdata.fda.gov/scripts/cder/ob/), Repurposed drug database (http://drugrepositioningportal.com/repurposed-drug-database.php), RepurposeDB (Hodos, Kidd et al. 2016), Drugs@FDA (https://www.accessdata.fda.gov/scripts/cder/daf/), NIH DailyMed (https://dailymed.nlm.nih.gov/dailymed/index.cfm), and EU Clinical Trials Register (https://www.clinicaltrialsregister.eu/ctr-search/). After obtaining drugs from these sources, the next step could be searching for drugs or compounds significantly similar to the

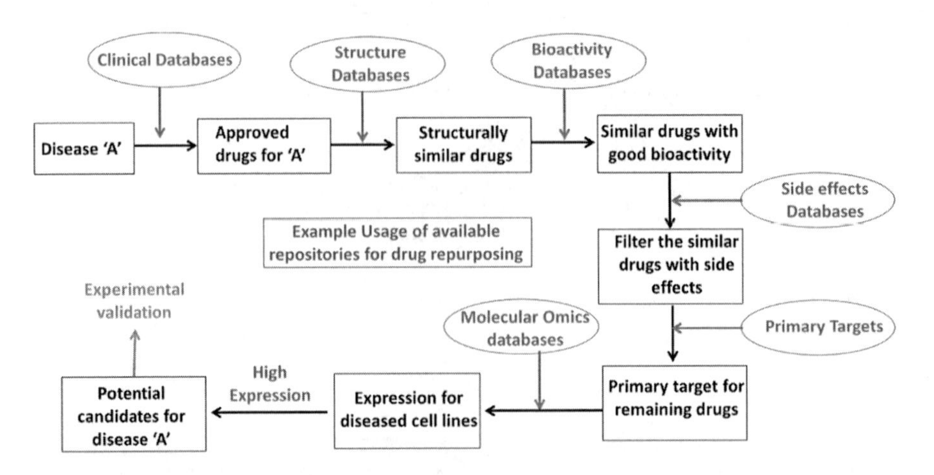

FIGURE 4.4 Example framework of disease-based drug repurposing with the usage of databases listed in Figure 4.3.

drugs for disease "A." The recommended chemical structure databases are PubChem (Wang, Bryant et al. 2017), ChemSpider (Kelly and Kidd2015), ChemDB (Chen, Swamidass et al. 2005), and DrugBank (Wishart, Knox et al. 2008). Next, to correlate the bioactivity between the drugs for disease "A" and a similar disease the Drug Target Profiler (DTP) (Tanoli, Alam et al. 2018a), Probes & Drugs portal (Skuta, Popr et al. 2017), BindingDB (Gilson, Liu et al. 2016), DrugCentral (Ursu, Holmes et al. 2019), Guide to PHARMACOLOGY (GtopDB) (Alexander, Kelly et al. 2017), and IDAAPM can be used. Once a significant correlation is noted, the side effects of similar drugs can be checked using databases under the category "drug side effects," that is, Sider (Kuhn, Letunic et al. 2016), VigiAccess (Shankar 2016), Integrated Database of ADMET, and Adverse Effects of Predictive Modeling (IDAAPM) (Legehar, Xhaard et al. 2016). Those similar drugs passing the side effects test can be further searched for the primary targets by using databases, namely, ChEMBL (Gaulton, Hersey et al. 2017) or DrugTargetCommons (DTC) (Tanoli, Alam et al. 2018b). Next, the identified protein target can be checked whether they are highly expressed in cell lines of disease "A" by accessing the databases, for example, Cancer Cell Line Encyclopedia (CCLE) (Barretina, Caponigro et al. 2012), GDSC (Iorio, Knijnenburg et al. 2016), or CellMinerCDB (Rajapakse, Luna et al. 2018). The potential new drugs, whose targets showed elevated expression, could be candidates for repurposing for disease "A" and can be experimentally validated.

4.1.3 Computational Approaches for Drug Repurposing

The combination of computational techniques with DR becomes "In Silico Repurposing" (Sohraby, Bagheri et al. 2019). The virtual screening and molecular dynamics (MD) simulation are well-known and effective methods to screen existing drugs for their effect on specific drug targets (Sahu et al., 2020; Rai et al., 2020). The molecular docking utilizes the scoring functions to calculate the binding of ligands to the active site of the receptors. Whereas, MD simulation is conducted to further examine the true positives and filter out the false positives from docking analysis (Sohraby, Bagheri et al. 2019). Here, we will discuss the methodology to find the most relevant drug target as well as a ligand to start with.

4.1.3.1 Literature Mining Approach

The literature embodies abundant knowledge of biology and medicine. The literature mining techniques are growing to enable access to the vast amount of information available in scientific articles, particularly to extract the relationships between biomedical entities. Advancement in methods of integration facilitates the construction of networks of associations between entities such as compounds, drugs, diseases, genes, and pathways (Jenssen, Laegreid et al. 2001; Chun, Tsuruoka et al. 2006). The available information for drugs and diseases can be extracted from the scientific literature by using text mining methodologies and these pieces of evidence further can be used for detecting potential indications of these drugs in repurposing.

The vast amount of information available in the literature can be used to infer new knowledge from the existing data (Deftereos, Andronis et al. 2011). The ABC model given by Swanson (1990) is a prominent methodology for literature-based discovery to generate a hypothesis between two concepts through a shared concept. The relationship between two scientific concepts **A** and **C** can be identified based on co-occurrence of concepts **A** and **B** in one publication and **B** and **C** in another publication.

The most commonly used repository for biomedical articles is PubMed (Macleod 2002) with a rapidly growing number of abstracts every year. The PubMed Central (2000) is a repository for full access to many articles. Together, they provide a large wealth of information for literature-based discoveries. For instance, if one article reports that disease **C** possesses a characteristic **B**, and in another publication, it is reported that substance **A** affects **B**, then we could discover a potential link between **A** and **C** based on common connections with **B**. The "closed discovery" and "open discovery" are two different processes based on the ABC model. Closed discovery refers to extracting the relationship between two already known concepts **A** and **C**. On the other hand, in open discovery, the search starts by searching for the concept **B** in connection with the concept **A** and then seeking for concept **C** connected with concept **B**. An example is searching for disease and its mechanism and then seeking a drug with the same mechanism of action.

The ABC model has been used to find indirect relationships between disease and drugs. (Yang, Ju et al. 2017). For example, celecoxib, a drug for osteoarthritis and rheumatoid arthritis, was found associated with ovarian cancer and breast cancer. The intermediates of both associations are vascular endothelial growth factor and cyclooxygenase-2. Celecoxib is a COX-2 inhibitor. The drug celecoxib in combination with paclitaxel may effectively treat ovarian cancer by regulating the paclitaxel-induced apoptosis through downregulation of NFkB (nuclear factor-kappa B) and Akt activation (Kim, Yim et al. 2014). Likewise, combining celecoxib and tamoxifen indicated an effective treatment of breast cancer via suppressing the expression of VEGF and VEGF receptor 2 (Kumar, Rajput et al. 2013). Several preclinical and clinical trials also demonstrated celecoxib as a suppressor of cancer cell proliferation and a potential candidate for combination therapy (Li, Hao et al. 2018). Various text mining tools are available with the aim to repurpose the drugs. These tools sufficiently extract the relationship between two or more different scientific entities. The tools for text mining are listed in Table 4.3.

4.1.3.2 Network-Based Approach

The idea of networking plays an important role in biological phenomena. These biological networks consist of genes, proteins, drugs, chemicals, and complexes (Alaimo and Pulvirenti 2019). The interconnections of these biological nodes further lead to generating novel hypotheses in DR. The biological network data can be accessed in various forms, namely, "gene-regulatory networks," "metabolic networks," "protein–protein interaction networks," and "drug interaction networks" (Gligorijevic and Przulj 2015; Chautard, Thierry-Mieg et al. 2009). For DR, this network data can be used either solely or by integrating with other data sources.

Any disease or drug administration can cause molecular perturbations, which can be observed by analyzing the gene expression data (Isik, Baldow et al. 2015; Cui, Zhang et al. 2018). Such information can be used to generate gene regulatory networks and also for gene prioritization and candidate gene identification for DR. In the literature, various mechanisms have been published on regulatory networks based on DR. Yeh, Yeh et al. (2012) have used an integrated network approach to find drug targets by incorporating the microarray data, disease–genes associations, and interactome network. Multiple protein–protein interactions were merged with regulatory factors. Another integrated network-based approach for DR was also conducted (Emig, Ivliev et al. 2013). They have integrated the microarray expression data (disease vs control) and a known regulatory network. The integrated network was analyzed by using four different algorithms, namely, "neighborhood scoring," "network propagation," "Random walk," and "interconnectivity." Potential candidates for breast and prostate cancer have been identified by using a network-based approach (Chen, Sherr et al. 2016). Their methodology was based on functional linkage network, where nodes represent genes or proteins and are reportedly interconnected based on common biological functions.

The metabolic networks are networks with nodes representing metabolites and chemical compounds and edges depict the reactions catalyzed by the enzymes. The excess concentration of a compound indicates the pathology (Alaimo and

TABLE 4.3
Appropriate Text Mining Tools for Drug Re-Purposing

Tool	Description	Web Link
Biovista	Gene–protein relationship	http://www.biovista.com/
EDGAR	Drug–gene relationships	https://www.sec.gov/
PolySearch	Drug–disease and drug–gene relationship	http://wishart.biology.ualberta.ca
DrugQuest	Drug–drug relationships	http://bioinformatics.med.uoc.gr
BEST	Extract the relevant bio-entities	http://best.korea.ac.kr.
BioWisdom	Relationship between drug–disease and drug–target	http://www.biowisdom.com

Pulvirenti 2019). These enzymes can be further explored as potential drug targets for DR. The metabolic networks are generally analyzed using flux balance analysis (FBA) (Orth,Thiele et al. 2010) to monitor the flow of metabolites. Folger, Jerby et al. (2011) have analyzed the metabolic network to detect alterations across different types of cancers. Li, Wang et al. (2011) have analyzed metabolic networks using FBA in the search for drug target identification.

The prediction of DTI can serve as a preliminary step for repurposing methodologies. Usually, a bipartite graph is used to represent the DTI network, wherein nodes signify either drug or targets (Alaimo and Pulvirenti 2019) and edges represent the validated interactions. DTI algorithm tends to identify the novel edges. It has been reported by Yamanishi, Araki et al. (2008) that drugs with similar structures could likely target the same protein, and target proteins with similar sequences are likely to interact with similar drugs (Yamanishi,Araki et al. 2008). Several other network-based approaches are available for DR. For instance, compounds with analogous chemical structures could act on similar proteins (Iorio,Bosotti et al. 2010). The knowledge of the side effects of drugs can be used to generate new hypotheses. As all drugs cause side effects due to off-targets, it may be considered for new uses (Gottlieb, Stein et al. 2011).

4.1.4 Web-Based Tools for Drug Repurposing

4.1.4.1 ACID

Auto in silico consensus inverse docking (ACID) is a user-friendly platform for DR and is freely accessible at http://chemyang.ccnu.edu.cn/ccb/server/ACID. It is based on consensus inverse docking protocol to enhance the success rate in comparison with single docking algorithms. It has been developed to address the challenges and technical limitations of drug-based approaches for DR. Inverse or reverse docking is the process of docking a single drug into possible binding sites with multiple disease targets, that is, "one ligand-many targets," and this approach has been proven very powerful for DR (Kharkar, Warrier et al. 2014). The consensus strategy includes different types of docking algorithms in the search for better pose predictions with a higher success rate. Hence, its development will overcome the inherent difficulties in molecular docking and likely serve for the identification of potential opportunities for DR (Lee, Lee et al. 2016). It stores information of about 2,764 commercial drugs (Figure 4.5).

4.1.4.2 repoDB

Nowadays, the computational repurposing is appealing as it offers to screen multiple drugs and proposes potential drug candidates, but is challenging due to inconsistent methods for validation (Figure 4.6). The repoDB (Drug Repurposing Database) (Brown and Patel 2017) is a publicly available repository of successful (true positives) and failed (true negative) repurposed drugs along with their indications. These drugs can further be considered for computational repurposing methods. The data in repoDB were drawn from two sources, namely, DrugCentral (Ursu,Holmes et al. 2017) and Clinical Trials Database (Tasneem,Aberle et al. 2012). The database search can be either "drugcentric" or "disease centric."

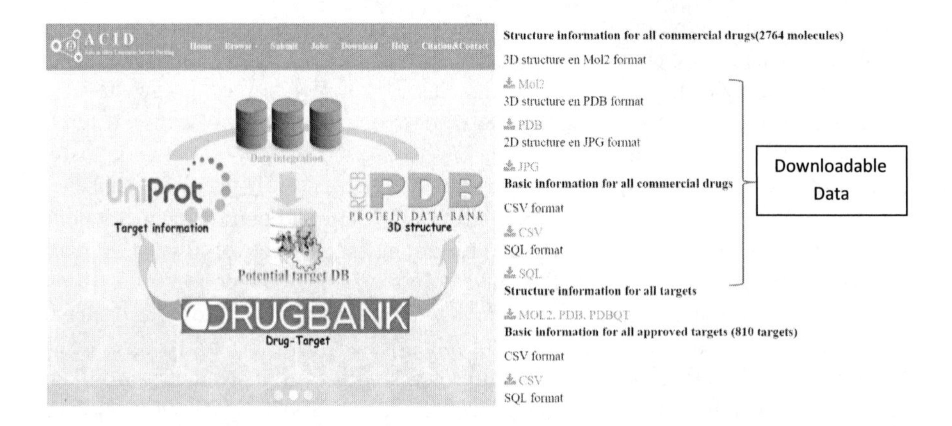

FIGURE 4.5 Auto in silico consensus inverse docking (ACID) homepage (left) and downloadable data (right).

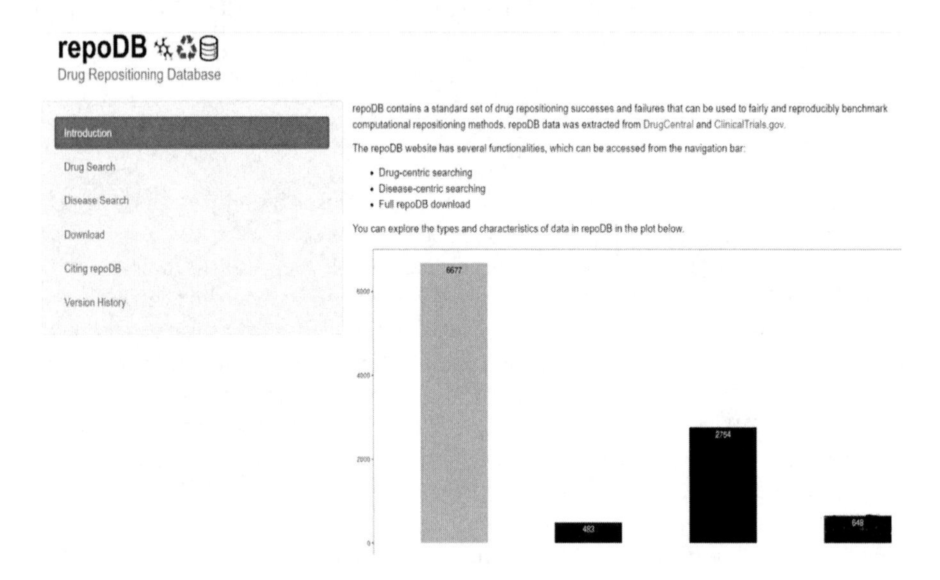

FIGURE 4.6 repoDB homepage. The search navigation bar on the left provides drug-centric and disease-centric search options. The graph represents the current statistics of the database.

It currently stores information of about 6,677 approved drugs, 483 suspended, 2,754 terminated, and 648 withdrawn drugs.

4.1.4.3 RE:FineDrugs

The RE:fine Drugs is a freely available platform that provides an integrated search as well as the discovery of drug candidates for repurposing (Moosavinasab, Patterson et al. 2016) by using GWAS and PheW as repurposing dataset. It aims to determine the potentials of identification and prioritization of novel drug candidates for repurposing based on drug–gene–disease triads. It provides for hypothesis generation rather than statistical testing, as its output proposes a possibility of DR only.

4.1.4.4 ReDO_DB

ReDO_DB stores a curated list of non-cancer drugs with anticancer activity (Pantziarka,Verbaanderd et al. 2018). The evidence is collected by analyzing the peer-reviewed studies, medical reports, clinical trials, and observational studies. After analyzing multiple reports, the resulting drug is further checked for its existence in the essential medicine list of the World Health Organization (WHO). The database is freely accessible at http://www.redo-project.org/db/.

4.1.4.5 RepurposeDB

It is a collection of repurposed drugs, their targets, and associated primary and secondary diseases (Jadamba and Shin 2016). The analysis of such data reveals the factors influencing DR. The enrichment analysis of the positively repurposed drugs in terms of pathways, physicochemical properties, mechanisms, pharmacological, and epidemiological factors could aid in designing better DR studies in the future. The developers performed a meta-analysis of RepurposeDB data and observed a need to modify the repurposing pipeline by including additional data and new strategies for analysis, namely, shared genetic architecture, pathway cross talks, and prevalence of disease comorbidities. Its source code is available at https://bitbucket.org/dudleylab/repurposedb

4.1.4.6 DeepDR

The network-based DR offered an extraordinary opportunity for the development of in silico approaches for DR (Zeng,Zhu et al. 2019). The DeepDR is a network-based deep learning approach for repurposing, which integrates 10 heterogeneous networks, namely, drug–disease network, drug–side effect network, drug–target network, and 7 different drug–drug networks. A multimodal deep auto encoder is used to learn the features of these diverse networks. The performance of DeepDR has been evaluated by using data from the ClinicalTrials.gov database as an external validation set. The authors have successfully identified drugs, for example, methylphenidate and pergolide for Parkinson's disease, and risperidone and aripiprazole for Alzheimer's disease. They have observed a 70% success rate after validating clinical studies and literature evidence. The source code and data are available at https://github.com/ChengF-Lab/deepDR.

4.1.4.7 Drug Vs Disease/DvD

The Drug versus Disease (Drug Vs Disease) is an R package (Pacini, Iorio et al. 2013), which offers the pipeline to compare the expression profiles of drug and diseased conditions. The resulting negative correlation between profiles may lead to generate hypotheses for DR; however, the positive correlation can be used to understand the side effects of drugs. It depends on the DvD data package, which contains the dataset of reference genomic profiles. The package can be directly installed from R Platform or the source package can be downloaded from https://www.bioconductor.org/packages/release/bioc/html/DrugVsDisease.html

4.1.5 Summary

In this chapter, we have highlighted the possible bioinformatics method for DR. The screening of all available approved drugs using bioinformatics techniques, such as molecular docking and MD simulation, is the broad way to start with DR. But the available bioinformatics web services precisely for repurposing further offer to reduce time and cost. These Web-based tools and databases aim to collect data required for DR. Different computational techniques can be used for the selection of drugs to be repurposed for new indications. The detailed analysis of drugs before screening for its reuse reduces time and enhances the probability of success in DR. Multiple categories of repositories publicly available for DR simplify the data access. The literature mining and networking skills can help in the identification of appropriate drug candidates for repurposing. The AI applications are also worth mentioning in the field of medicine. Food and Drug Administration (FDA) has also approved some AI-based algorithms (Briganti and Le Moine 2020). Such technologies enable improving the clinical practice thereby opening a new branch in medicine. The above described computational approaches work on a given aspect of the disease, whereas the AI-based technologies have the potential to integrate multiple data thereby increasing efficiency. Therefore, computational methods and bioinformatics approaches offer rapid ways to obtain a starting point for repurposing.

REFERENCES

Alaimo, S. and A. Pulvirenti (2019). "Network-based drug repositioning: approaches, resources, and research directions." *Methods Mol Biol* 1903: 97–113.

Alexander, S. P., E. Kelly, N. V. Marrion, J. A. Peters, E. Faccenda, S. D. Harding, A. J. Pawson, J. L. Sharman, C. Southan, O. P. Buneman, J. A. Cidlowski, A. Christopoulos, A. P. Davenport, D. Fabbro, M. Spedding, J. Striessnig, J. A. Davies and C. Collaborators (2017). "The concise guide to pharmacology 2017/18: overview." *Br J Pharmacol*174 Suppl 1: S1–S16.

Anderson, K. C. (2001). "Dark remedy: the impact of Thalidomide and its revival as a vital medicine." *Nature Medicine* 7(3): 275–276.

Ashburn, T. T. and K. B. Thor (2004). "Drug repositioning: identifying and developing new uses for existing drugs." *Nat Rev Drug Discov* 3(8): 673–683.

Barretina, J., G. Caponigro, N. Stransky, K. Venkatesan, A. A. Margolin, S. Kim, C. J. Wilson, J. Lehar, G. V. Kryukov, D. Sonkin, A. Reddy, M. Liu, L. Murray, M. F. Berger, J. E. Monahan, P. Morais, J. Meltzer, A. Korejwa, J. Jane-Valbuena, F. A. Mapa, J. Thibault, E. Bric-Furlong, P. Raman, A. Shipway, I. H. Engels, J. Cheng, G. K. Yu, J. Yu, P. Aspesi, Jr., M. de Silva, K. Jagtap, M. D. Jones, L. Wang, C. Hatton, E. Palescandolo, S. Gupta, S. Mahan, C. Sougnez, R. C. Onofrio, T. Liefeld, L. MacConaill, W. Winckler, M. Reich, N. Li, J. P. Mesirov, S. B. Gabriel, G. Getz, K. Ardlie, V. Chan, V. E. Myer, B. L. Weber, J. Porter, M. Warmuth, P. Finan, J. L. Harris, M. Meyerson, T. R. Golub, M. P. Morrissey, W. R. Sellers, R. Schlegel and L. A. Garraway (2012). "The cancer cell line encyclopedia enables predictive modelling of anticancer drug sensitivity." *Nature* 483(7391): 603–607.

Bellomo, F., D. L. Medina, E. De Leo, A. Panarella and F. Emma (2017). "High-content drug screening for rare diseases." *J Inherit Metab Dis* 40(4): 601–607.

Boolell, M., M. J. Allen, S. A. Ballard, S. Gepi-Attee, G. J. Muirhead, A. M. Naylor, I. H. Osterloh and C. Gingell (1996). "Sildenafil: an orally active type 5 cyclic GMP-specific phosphodiesterase inhibitor for the treatment of penile erectile dysfunction."*Int J Impot Res* 8(2): 47–52.

Breckenridge, A. and R. Jacob (2019). "Overcoming the legal and regulatory barriers to drug repurposing." *Nat Rev Drug Discov* 18(1): 1–2.

Briganti, G. and O. Le Moine (2020). "Artificial intelligence in medicine: today and tomorrow." *Front Med (Lausanne)* 7: 27.

Brown, A. S. and C. J. Patel (2017). "A standard database for drug repositioning." *Sci Data* 4: 170029.

Chautard, E., N. Thierry-Mieg and S. Ricard-Blum (2009). "Interaction networks: from protein functions to drug discovery. A review." *PatholBiol (Paris)* 57(4): 324–333.

Chen, H. R., D. H. Sherr, Z.Hu and C. DeLisi (2016). "A network based approach to drug repositioning identifies plausible candidates for breast cancer and prostate cancer." *BMC Med Genomics* 9(1): 51.

Chen, J., S. J. Swamidass, Y. Dou, J. Bruand and P. Baldi (2005). "ChemDB: a public database of small molecules and related chemoinformatics resources." *Bioinformatics* 21(22): 4133–4139.

Chen, N., S. Zhou and M. Palmisano (2017). "Clinical pharmacokinetics and pharmacodynamics of lenalidomide."*ClinPharmacokinet* 56(2): 139–152.

Chun, H. W., Y. Tsuruoka, J. D. Kim, R. Shiba, N. Nagata, T. Hishiki and J. Tsujii (2006). "Extraction of gene-disease relations from Medline using domain dictionaries and machine learning." *Pac SympBiocomput* 2006: 4–15.

Cui, H., M. Zhang, Q. Yang, X. Li, M. Liebman, Y. Yu and L. Xie (2018)."The prediction of drug-disease correlation based on gene expression data." *Biomed Res Int* 2018: 4028473.

de Oliveira, E. A. M. and K. L. Lang (2018). "Drug repositioning: concept, classification, methodology, and importance in rare/orphans and neglected diseases." *Journal of Applied Pharmaceutical Science* 8(08): 157–165.

Deftereos, S. N., C. Andronis, E. J. Friedla, A. Persidis and A. Persidis (2011). "Drug repurposing and adverse event prediction using high-throughput literature analysis." *Wiley Interdiscip Rev SystBiol Med* 3(3): 323–334.

DiMasi, J. A., L. Feldman, A. Seckler and A. Wilson (2010). "Trends in risks associated with new drug development: success rates for investigational drugs." *ClinPharmacolTher* 87(3): 272–277.

DiMasi, J. A., H. G. Grabowski and R. W. Hansen (2016). "Innovation in the pharmaceutical industry: New estimates of R&D costs." *J Health Econ*47: 20–33.

Dudley, J. T., T. Deshpande and A. J. Butte (2011). "Exploiting drug-disease relationships for computational drug repositioning." *Brief Bioinform* 12(4): 303–311.

Emig, D., A. Ivliev, O. Pustovalova, L. Lancashire, S. Bureeva, Y. Nikolsky and M. Bessarabova (2013). "Drug target prediction and repositioning using an integrated network-based approach." *PLoS One* 8(4): e60618.

Folger, O., L. Jerby, C. Frezza, E. Gottlieb, E. Ruppin and T. Shlomi (2011). "Predicting selective drug targets in cancer through metabolic networks."*Mol SystBiol*7: 501.

Gaulton, A., A. Hersey, M. Nowotka, A. P. Bento, J. Chambers, D. Mendez, P. Mutowo, F. Atkinson, L. J. Bellis, E. Cibrian-Uhalte, M. Davies, N. Dedman, A. Karlsson, M. P. Magarinos, J. P. Overington, G. Papadatos, I. Smit and A. R. Leach (2017). "The ChEMBL database in 2017."*Nucleic Acids Res* 45(D1): D945–D954.

Gilson, M. K., T. Liu, M. Baitaluk, G. Nicola, L. Hwang and J. Chong (2016). "BindingDB in 2015: A public database for medicinal chemistry, computational chemistry and systems pharmacology." *Nucleic Acids Res* 44(D1): D1045–D1053.

Gligorijevic, V. and N.Przulj (2015). "Methods for biological data integration: Perspectives and challenges."*J R Soc Interface*12(112).

Gottlieb, A., G. Y. Stein, E. Ruppin and R. Sharan (2011). "PREDICT: a method for inferring novel drug indications with application to personalized medicine." *Mol SystBiol* 7: 496.

Hatzimouratidis, K. (2006). "Sildenafil in the treatment of erectile dysfunction: An overview of the clinical evidence."*ClinInterv Aging* 1(4): 403–414.

Hodos, R. A., B. A. Kidd, K. Shameer, B. P. Readhead and J. T. Dudley (2016). "In silico methods for drug repurposing and pharmacology." *Wiley Interdiscip Rev SystBiol Med* 8(3): 186–210.

Iorio, F., R. Bosotti, E. Scacheri, V. Belcastro, P. Mithbaokar, R. Ferriero, L. Murino, R. Tagliaferri, N. Brunetti-Pierri, A. Isacchi and D. di Bernardo (2010). "Discovery of drug mode of action and drug repositioning from transcriptional responses." *Proc Natl Acad Sci U S A* 107(33): 14621–14626.

Iorio, F., T. A. Knijnenburg, D. J. Vis, G. R. Bignell, M. P. Menden, M. Schubert, N. Aben, E. Goncalves, S. Barthorpe, H. Lightfoot, T. Cokelaer, P. Greninger, E. van Dyk, H. Chang, H. de Silva, H. Heyn, X. Deng, R. K. Egan, Q. Liu, T. Mironenko, X. Mitropoulos, L. Richardson, J. Wang, T. Zhang, S. Moran, S. Sayols, M. Soleimani, D. Tamborero, N. Lopez-Bigas, P. Ross-Macdonald, M. Esteller, N. S. Gray, D. A. Haber, M. R. Stratton, C. H.Benes, L. F. A. Wessels, J. Saez-Rodriguez, U. McDermott and M. J. Garnett (2016). "A landscape of pharmacogenomic interactions in cancer." *Cell* 166(3): 740–754.

Isik, Z., C. Baldow, C. V. Cannistraci and M. Schroeder (2015)."Drug target prioritization by perturbed gene expression and network information." *Sci Rep*5: 17417.

Jadamba, E. and M. Shin (2016)."A systematic framework for drug repositioning from integrated omics and drug phenotype profiles using pathway-drug network." *Biomed Res Int* 2016: 7147039.

Jenssen, T. K., A. Laegreid, J. Komorowski and E. Hovig (2001). "A literature network of human genes for high-throughput analysis of gene expression." *Nat Genet* 28(1): 21–28.

Jin, G. and S. T. Wong (2014). "Toward better drug repositioning: prioritizing and integrating existing methods into efficient pipelines." *Drug Discov Today*19(5): 637–644.

Kelly, R., and Kidd, R. (2015).Editorial: ChemSpider—A tool for natural products research. *Nat Prod Rep* 32(8), 1163–1164.

Kharkar, P. S., S. Warrier and R. S. Gaud (2014). "Reverse docking: a powerful tool for drug repositioning and drug rescue."*Future Med Chem* 6(3): 333–342.

Khatoon, N., H.Alam, A. Khan, K. Raza and M. Sardar (2019). "Ampicillin silver nanoformulations against multidrug resistant bacteria."*Sci Rep* 9(1): 6848.

Kim, H. J., G. W. Yim, E. J. Nam and Y. T. Kim (2014). "Synergistic effect of COX-2 inhibitor on paclitaxel-induced apoptosis in the human ovarian cancer cell line OVCAR-3." *Cancer Res Treat* 46(1): 81–92.

Koch, U., M. Hamacher and P. Nussbaumer (2014). "Cheminformatics at the interface of medicinal chemistry and proteomics."*BiochimBiophysActa* 1844(1 PtA): 156–161.

Kuhn, M., I. Letunic, L. J. Jensen and P. Bork (2016). "The SIDER database of drugs and side effects." *Nucleic Acids Res* 44(D1): D1075–D1079.

Kumar, B. N., S. Rajput, K. K. Dey, A. Parekh, S. Das, A. Mazumdar and M. Mandal (2013). "Celecoxib alleviates tamoxifen-instigated angiogenic effects by ROS-dependent VEGF/VEGFR2 autocrine signaling." *BMC Cancer* 13: 273.

Langedijk, J., A. K. Mantel-Teeuwisse, D. S. Slijkerman and M. H. Schutjens (2015)."Drug repositioning and repurposing: terminology and definitions in literature." *Drug Discov Today* 20(8): 1027–1034.

Lee, A., K. Lee and D. Kim (2016). "Using reverse docking for target identification and its applications for drug discovery." *Expert Opin Drug Discov* 11(7): 707–715.

Legehar, A., H. Xhaard and L. Ghemtio (2016). "IDAAPM: integrated database of ADMET and adverse effects of predictive modeling based on FDA approved drug data." *J Cheminform*8: 33.

Li, J., Q. Hao, W. Cao, J. V. Vadgama and Y. Wu (2018)."Celecoxib in breast cancer prevention and therapy." *Cancer Manag Res* 10: 4653–4667.

Li, Y. Y. and S. J. Jones (2012). "Drug repositioning for personalized medicine."*Genome Med* 4(3): 27.

Li, Z., R. S. Wang and X. S. Zhang (2011). "Two-stage flux balance analysis of metabolic networks for drug target identification." *BMC SystBiol*5 Suppl 1: S11.

Macleod, M. R. (2002). "PubMed: http://www.pubmed.org." *J NeurolNeurosurg Psychiatry* 73(6): 746.

Moffat, J. G., F. Vincent, J. A. Lee, J. Eder and M. Prunotto (2017). "Opportunities and challenges in phenotypic drug discovery: an industry perspective." *Nat Rev Drug Discov* 16(8): 531–543.

Mohs, R. C. and N. H. Greig (2017). "Drug discovery and development: Role of basic biological research." *Alzheimers Dement (N Y)* 3(4): 651–657.

Moosavinasab, S., J. Patterson, R. Strouse, M. Rastegar-Mojarad, K. Regan, P. R. Payne, Y. Huang and S. M. Lin (2016). "'RE:fine drugs": an interactive dashboard to access drug repurposing opportunities." *Database (Oxford)* 2016: baw083.

Murteira, S., Z. Ghezaiel, S. Karray and M. Lamure (2013). "Drug reformulations and repositioning in pharmaceutical industry and its impact onmarket access: reassessment of nomenclature." *J Mark Access Health Policy* 1(1): 21131.

Oprea, T. I. and J. P. Overington (2015)."Computational and practical aspects of drug repositioning." *Assay Drug DevTechnol* 13(6): 299–306.

Orth, J. D., I. Thiele and B. O. Palsson (2010)."What is flux balance analysis?" *Nat Biotechnol* 28(3): 245–248.

Pacini, C., F. Iorio, E. Goncalves, M. Iskar, T. Klabunde, P. Bork and J. Saez-Rodriguez (2013). "DvD: An R/Cytoscape pipeline for drug repurposing using public repositories of gene expression data." *Bioinformatics* 29(1): 132–134.

Pantziarka, P., C. Verbaanderd, V. Sukhatme, I. Rica Capistrano, S. Crispino, B. Gyawali, I. Rooman, A. M. Van Nuffel, L. Meheus, V. P. Sukhatme and G.Bouche (2018). "ReDO_DB: the repurposing drugs in oncology database." *Ecancermedicalscience* 12: 886.

PubMed Central (2000). "Signing on." *CMAJ* 162(4): 481, 483.

Pushpakom, S., F. Iorio, P. A. Eyers, K. J. Escott, S. Hopper, A. Wells, A. Doig, T. Guilliams, J. Latimer, C. McNamee, A. Norris, P. Sanseau, D. Cavalla and M. Pirmohamed (2019). "Drug repurposing: progress, challenges and recommendations." *Nat Rev Drug Discov* 18(1): 41–58.

Rai, A., S. Qazi and K. Raza (2020). In silico analysis and comparative molecular docking study of FDA approved drugs with Transforming Growth Factor Beta receptors in Oral Submucous Fibrosis. *Indian J Otolaryngol., Head Neck Surg.*, Springer. https://doi.org/10.1007/s12070-020-02014-5.

Rajapakse, V. N., A. Luna, M. Yamade, L. Loman, S. Varma, M. Sunshine, F. Iorio, F. G. Sousa, F. Elloumi, M. I. Aladjem, A. Thomas, C. Sander, K. W. Kohn, C. H. Benes, M. Garnett, W. C. Reinhold and Y. Pommier (2018). "CellMinerCDB for integrative cross-database genomics and pharmacogenomics analyses of cancer cell lines." *iScience* 10: 247–264.

Readhead, B. and J. Dudley (2013). "Translational bioinformatics approaches to drug development." *Adv Wound Care (New Rochelle)* 2(9): 470–489.

Sahu, A., S. Qazi., K. Raza and S. Varma (2020). COVID-19: *Hard Road to Find Integrated Computational Drug and Repurposing Pipeline. Computational Intelligence Methods in COVID 19: Surveillance, Prevention, Prediction and Diagnosis*, Studies in Computational Intelligence (SCI), Springer, 923: 295-309. https://doi.org/10.1007/978-981-15-8534-0_15.

Sam, E. and P. Athri (2019). "Web-based drug repurposing tools: A survey." *Brief Bioinform* 20(1): 299–316.

Sampaio, E. P., E. N. Sarno, R. Galilly, Z. A. Cohn and G. Kaplan (1991). "Thalidomide selectively inhibits tumor necrosis factor alpha production by stimulated human monocytes." *J Exp Med* 173(3): 699–703.

Shankar, P. R. (2016). "VigiAccess: Promoting public access to VigiBase." *Indian J Pharmacol* 48(5): 606–607.

Skuta, C., M.Popr, T. Muller, J. Jindrich, M. Kahle, D. Sedlak, D. Svozil and P. Bartunek (2017). "Probes &Drugs portal: an interactive, open data resource for chemical biology." *Nat Methods* 14(8): 759–760.

Sohraby, F., M. Bagheri and H. Aryapour (2019). "Performing an in silico repurposing of existing drugs by combining virtual screening and molecular dynamics simulation." *Methods Mol Biol* 1903: 23–43.

Sun, W., P. E. Sanderson and W. Zheng (2016). "Drug combination therapy increases successful drug repositioning." *Drug Discov Today* 21(7): 1189–1195.

Swanson, D. R. (1990). "Medical literature as a potential source of new knowledge." *Bull Med LibrAssoc* 78(1): 29–37.

Tanoli, Z., Z. Alam, A. Ianevski, K. Wennerberg, M. Vaha-Koskela and T. Aittokallio (2018a). "Interactive visual analysis of drug-target interaction networks using Drug Target Profiler, with applications to precision medicine and drug repurposing." *Brief Bioinform* 21(1): 211–220.

Tanoli, Z., Z. Alam, M. Vaha-Koskela, B. Ravikumar, A. Malyutina, A. Jaiswal, J. Tang, K. Wennerberg and T. Aittokallio (2018b). "Drug Target Commons 2.0: A community platform for systematic analysis of drug-target interaction profiles." *Database (Oxford)* 2018: 1–13.

Tanoli, Z., U. Seemab, A. Scherer, K. Wennerberg, J. Tang and M. Vaha-Koskela (2020). "Exploration of databases and methods supporting drug repurposing: A comprehensive survey." *Brief Bioinform* bbaa003

Tasneem, A., L. Aberle, H. Ananth, S. Chakraborty, K. Chiswell, B. J. McCourt and R. Pietrobon (2012). "The database for aggregate analysis of ClinicalTrials.gov (AACT) and subsequent regrouping by clinical specialty." *PLoS One* 7(3): e33677.

Urquhart, L. (2018). "Market watch: Top drugs and companies by sales in 2017." *Nat Rev Drug Discov* 17(4): 232.

Ursu, O., J. Holmes, C. G. Bologa, J. J. Yang, S. L. Mathias, V. Stathias, D. T. Nguyen, S. Schurer and T. Oprea (2019). "DrugCentral 2018: An update." *Nucleic Acids Res* 47(D1): D963–D970.

Ursu, O., J. Holmes, J. Knockel, C. G. Bologa, J. J. Yang, S. L. Mathias, S. J. Nelson and T. I. Oprea (2017). "DrugCentral: Online drug compendium." *Nucleic Acids Res* 45 (D1): D932–D939.

Wang, Y., S. H. Bryant, T. Cheng, J. Wang, A. Gindulyte, B. A. Shoemaker, P. A. Thiessen, S.He and J. Zhang (2017). "PubChemBioAssay: 2017 update." *Nucleic Acids Res* 45(D1): D955–D963.

Wishart, D. S., C. Knox, A. C. Guo, D. Cheng, S. Shrivastava, D. Tzur, B. Gautam and M. Hassanali (2008). "DrugBank: a knowledgebase for drugs, drug actions and drug targets." *Nucleic Acids Res* 36 (Database issue): D901–906.

Wu, H., J. Huang, Y. Zhong and Q. Huang (2017). "DrugSig: A resource for computational drug repositioning utilizing gene expression signatures."*PLoS One* 12(5): e0177743.

Xue, H., J. Li, H. Xie and Y. Wang (2018). "Review of drug repositioning approaches and resources." *Int J BiolSci* 14(10): 1232–1244.

Yamanishi, Y., M. Araki, A. Gutteridge, W. Honda and M. Kanehisa (2008)."Prediction of drug-target interaction networks from the integration of chemical and genomic spaces." *Bioinformatics* 24(13): i232–240.

Yang, H. T., J. H. Ju, Y. T. Wong, I. Shmulevich and J. H. Chiang (2017). "Literature-based discovery of new candidates for drug repurposing." *Brief Bioinform* 18(3): 488–497.

Yang, X., Y. Wang, R. Byrne, G. Schneider and S. Yang (2019). "Concepts of artificial intelligence for computer-assisted drug discovery." *Chem Rev* 119(18): 10520–10594.

Yeh, S. H., H. Y. Yeh and V. W. Soo (2012). "A network flow approach to predict drug targets from microarray data, disease genes and interactome network - case study on prostate cancer." *J ClinBioinforma* 2(1): 1.

Zeng, X., S. Zhu, X. Liu, Y. Zhou, R. Nussinov and F. Cheng (2019). "deepDR: A network-based deep learning approach to in silico drug repositioning." *Bioinformatics* 35(24): 5191–5198.

Part II

Internet of Things, Viroinformatics, and Toxin Databases for Healthcare Applications

5 The Fundamentals and Potential of IoT for Bioinformatics and Healthcare

Reinaldo Padilha França, Ana Carolina Borges Monteiro, Rangel Arthur and Yuzo Iano
State University of Campinas (UNICAMP)

CONTENTS

5.1 INTRODUCTION

The Internet of Things (IoT) is the concept that describes the universe of objects connected to the Internet and their attributions. In bioinformatics, this technology can be found in wearables, and connected medical equipment are some examples of the use of IoT to capture data and optimize the routine of professionals and patients [1,2].

IoT is an important asset for the entire value chain in the sector, benefiting patients, hospitals, operators, research and development institutes, professionals, and pharmaceutical laboratories. The technology through bioinformatics allows communication between equipment and improves medical carethrough preventive diagnosis of diseases, thereby helping to save lives [2,3].

The advantages of IoT in health have connected devices that bring together several advantages such as autonomous information recording, continuous patient monitoring, ease of data sharing, greater access to health information, automatic storage in the cloud, more complete medical history, with support to assertive diagnoses, patient empowerment, strengthening preventive and self-care actions, among others [4].

The eHealth through IoT technology can act on several fronts in medicine, such as keeping patients more connected to doctors through remote monitoring and virtual visits, helping hospitals to track employees and patients, facilitating care and monitoring of chronic diseases, automating the patient care workflow, reducing inefficiency and errors, optimizing the pharmaceutical manufacturing process, maintaining quality control and managing sensitive items while they are in transit, and lowering healthcare costs by simplifying the overall process, among others[5].

Several devices can be connected to the Wi-Fi network or others in a hospital, generating large amounts of data and allowing the tracking of information. Professionals, teams, and patients can benefit from IoT, with their health activities and conditions monitored. In the case of professionals, their stress levels can be monitored, aiming to reduce factors such as pressure and work overload, in addition to suggesting favorable moments for making important decisions, increasing the safety of these actions. Patients can be monitored using medical equipment that sends messages to the nursing staff or specialists on call when necessary [6].

The IoT in healthcare and medicine has had different roles, helping hospitals, doctors, and other professionals in the field both in their daily routine and in more complex functions, such as the treatment of certain types of cancer. One application of this system is in the area of bioinformatics, as this science makes use of statistical data to model the behavior of certain biological environments over time [6].

Therefore, this chapter has the mission and objective of providing an updated review and overview of IoT applied in healthcare, addressing its evolution and fundamental concepts, showing its relationship with bioinformatics as well as approaching its success, with a concise bibliographic background, categorizing and synthesizing the potential of the technique.

5.2 TRANSLATIONAL MEDICINE

Traditionally, biomedical research is divided into two watertight groups: basic research and clinical research. In most cases, there is no connection between the two. With an academic dichotomy between basic and clinical research predominating, there is a permanent gap between these two types of research. For this reason, the knowledge produced by basic research is not well used for practical purposes or, at best, it is used very slowly and with the little promise [7].

Translational research was born from these "promises": first, to unravel the bottlenecks of the technological innovation process in new medicines and therapies from a new perspective of problem analysis, the need to address the gap initially identified between the basic researcher and the clinical researcher, that is, between science-based knowledge and innovation in the health field, creating "bridges" between research and practice; second, to bring basic research closer to clinical research in a bidirectional and integrated way, which has shown relevance as a potential instrument to accelerate technological innovation and to approximate basic research results with effective applications in meeting demands of health [8,9].

Translational research in medicine begins with the identification of the disease that affects the patient and the knowledge of its pathophysiology for the search for possible therapeutic targets, that is, from the patient's bed to the laboratory bench. Consequently, potential treatments are developed based on basic laboratory research, and subsequently, the safety and efficacy of these drugs are tested through clinical trials, that is, from the laboratory bench to the patient's bed. Therefore, aiming to expedite the transfer of results from basic research to clinical research, to produce benefits for society. Consequently, the success of this new strategy has enabled the discovery of new drugs and their development in products for human health [8,9].

Translational medicine is associated with the transfer of biomedical knowledge generated in basic research to different areas of clinical research, especially concerning the diagnosis, treatment, and prevention of diseases, to promote integration between researchers and users of scientific research. It is the science in which scientists, doctors, and health professionals work together in search of innovations that allow improving diagnoses and treatments [10,11].

Basic science can contribute very significantly to the improvement of clinical care outcomes. Besides, there is great potential for collaboration between the clinical and research areas of our institution to promote health innovation. In this sense, the goal of translational medicine is to take the discoveries of science to the care practice and feedback research based on the main challenges that doctors encounter in their daily lives. This two-way street, in which scientists, doctors, and other health professionals work together, seeking innovations for the diagnosis and treatment of diseases, and thus promoting the dialogue between the two extremes of medicine—basic science and clinic—is a starting point for creating new projects [10,11].

Translational medicine is defined as an interdisciplinary branch of the field of biomedical knowledge that is based on three pillars related to bench, bedside, and community, and aims to combine disciplines, resources, techniques, and expertise to promote advances in prevention, diagnosis, and treatment of diseases, acting as a process that brings basic science and clinical application closer together, toward sustainable solutions to community health problems [12].

5.3 BIOTECHNOLOGY FOR TRANSLATIONAL RESEARCH WITH A FOCUS ON HUMAN HEALTHCARE

Innovation in biotechnology focused on human health is directly related to the discovery of new drugs. In this way, research is the support for the development of new practices in the production environment, as well as in the generation of new and better products and services. The gap between the beginning of the research process and its end is wide and indeterminate, and some research developments can cross generations of individuals, and to avoid this, in the complex area of healthcare, one of the main research development models is the translational model, which requires a strong and intensive intellectual property multidisciplinary strategy [13].

Considering that it is the fastest way for scientific information to be transmitted both for clinical research and for care practice, meaning an effective transfer of new knowledge, mechanisms, and techniques for the health of the population, collaborating, in an unequivocal manner, for medical decision-making refers to the transfer, exchange, and interpretation of data obtained in both basic and medical research in the clinical area [13].

The findings from laboratory-research and preclinical studies are applied for the development of clinical trials and studies in humans. Its interdisciplinary approach is oriented toward a kind of "reengineering" of the flow of development of new drugs, with the sense of being a potential instrument to accelerate technological innovation in the health area. The second area of concern for translational research reinforces the adoption of best practices [13,10].

The relevance of translational research is not in its object itself but in the emphasis dedicated to the urgent need to identify and solve the problems that undermine the effective transfer of scientific advancement in applied and useful knowledge. Since it goes through the exploration of a new perspective of analysis that integrally sees the development flow (considering all its amplitude), it expands to the whole set of existing activities between the discovery resulting from basic research and its conversion, into an application for the population in medical practice [11].

The incorporation of these stages of pre- and post-technological development occurs from the evolution of the understanding of the translational process from a systemic perspective of health innovation, from discovery to the patient. A context of overspecialization and consequent fragmentation of biomedical research activities is emerging, which brings the need for greater communication between the basic and clinical areas [10].

In this concept, the development of translational medicine aims to be applied to the benefit of society, necessarily implying the effort of developers to carry out in a single arc the entire spectrum of research. Thus, it proposes to fill this gap between the basic science researcher and the clinician in their fields of practice. This is understood as a set of efforts used to transfer the results obtained from basic research to adoption and consumption by users/patients, to meet a health need, and as a premise the permanent interdependence and integration between the stages of the development flow. Thus, in research that is developed with translational characteristics, its primary objective is to achieve innovation [8].

The classic stages of the technological development process, "from the bench to the bed," are considered as part of the scope of translational medicine, including studies of basic research to discover new molecules focusing on their transition to safe clinical applications. Thus, the research makes it possible to expand access to new treatments, improve quality of life, and increase patient survival. Translational medicine brings knowledge from laboratories to clinical practice, and from clinical practice, many ideas emerge for new research [10].

5.4 TRANSLATIONAL MEDICINE'S FOCUS ON INTERDISCIPLINARY GAP

Thus, to minimize this interdisciplinary gap and support translational medicine, providing the articulation between the laboratory, where the discoveries of basic science are developed, and the clinic, where practical applications are carried out, the multidisciplinary guidelines are followed bringing together researchers from various fields of knowledge, aiming at the multidisciplinary integration between different areas of common interest, and in this way, rationalize the technical and scientific capacity, as well as the time necessary to act in translational research, contributing to new types of disease diagnosis and acting in the development of new drugs [14].

The relevance of research and translational medicine resides in the approximation of the areas of knowledge of exact sciences such as mathematics, informatics, and engineering with the biological areas of biomedicine and medicine, in terms of reciprocal learning and its consequent advances in the literature for the treatment of problems related to both fields. With the motivation to treat the "gaps" of the development flow of medicines and new therapies and/or medicines, it should focus on fulfilling the promise that translational research should improve the health and longevity of the world's populations [11].

To achieve this, it is important to emphasize that research and development of new drugs involves a multidisciplinary study, at the same time translational researchers must be prepared to unite the discoveries of the basic sciences with the broad territory of clinical research and translate these results into modifications of the medical practice [15].

Medicine and other areas of health are shown as sciences in full evolution, resulting in a paradigm shift, where most research has ceased to be merely the description of a given physiological state/disease and has started to seek an understanding of what goes on in the cell and its molecular and metabolic pathways. The innovative technical and scientific potential of translational medicine working in the different phases of the development flow should decisively contribute to the development of translational medicine, from the patient's bed to the laboratory bench, from the bedside to bench, and from the laboratory to the patient's bed, "from bench to bedside" [9,14].

5.5 IoT IN BIOINFORMATICS AND BIOTECHNOLOGY

With the large volume of data available today, it is difficult to find relevant content. The IoT consists of a world of physical objects embedded with sensors and actuators connected by wireless networks and communicating using the Internet, shaping a network of intelligent objects, with processing capacity and capable of capturing environmental variables and reacting to external stimuli. These objects are connected and can be controlled over the Internet, enabling a multitude of new applications [16].

The number of devices connected to the IoT increases every day. The advancement of technology entering our daily lives is increasing, contributing to the flexibility and mobility of the most diverse tasks: security, home automation, industry 4.0, hospital automation, and medical services, among others. This is the concept of IoT in medicine where all computerized devices can interconnect, using the Internet [17].

Although the Web has been used extensively in the last decade, new technological standards have emerged as web services, the composition of these web services implies a new form of data management, concerning data in XML format, new management techniques, transactions, and XML query processing. What it implies from the point of view of information systems, considering the cheapening of computer systems, combined with software models for sharing distributed memory, is that the storage and manipulation of medical data in hospitals and clinics with parallel processing is facilitated, therefore providing environments that integrate varied computational resources, which are managed by different organizations and geographically distributed, bringing an innovative need in the management of this data [18].

IoT is one of the main technologies to allow the creation of cyber-physical systems and to realize intelligent vision from this scenario. Several recent technological advances have allowed the emergence of IoT applied to bioinformatics, such as nanotechnology, wireless sensor networks, mobile communication, and ubiquitous computing. However, there are still technical challenges to be overcome to fully realize the IoT paradigm [19].

In this sense, bioinformatics applications typically manage a large volume of data, which are multidimensional, dynamic, with different levels of complexity, and coming from several heterogeneous sources such as sensory data, protein and gene sequencing, and image digitization, among many others. The technological challenges are related to the design of solutions, such as middleware platforms, dealing with this enormous heterogeneity resulting from the diversity of hardware, sensors and actuators, and wireless technologies inherent to IoT bioinformatics [3].

The development of IoT applications in health considers the scale and heterogeneity of medical and biomedical devices in the healthcare environment. The processing and storage of the huge amount of data generated, often in the form of streams that require online computing, as also the management of the resources, generally heterogeneous, is necessary to handle with this data to provide answers and information of added value and on time for bioinformatics applications. Such characteristics demand the use of advanced technology for adequate management and satisfactory performance in the manipulation of the stored data, extraction, and management of knowledge from that data [3,20–22].

Bioinformatics is one of the results generated from that of Watson and Crick in 1953, which revealed that the DNA is structured as a double helix. The researchers could not imagine the volume of information that would exponentially be generated from that moment on. In the decades that followed the work of Watson and Crick, computational tools, which were initially quite simplified, began to be developed and made it possible to analyze and resolve numerous questions related to the structure of DNA, as well as the genetic information that encodes proteins, their structural properties, and the factors that regulate them, as well as the events associated with genetic regulation, with the molecular bases of embryonic development and with the evolution of metabolic and biochemical pathways. Simply put, bioinformatics is the union of computer techniques with molecular biology, that is, it is an area that needs professionals linked with multiplicity [23–25].

Since the 1960s, the growth in the number of known amino-acid sequences has led to the pioneering application of computers in molecular biology. From that period, the amount of data that should be analyzed has grown considerably, and the more affordable prices of computers have made it possible to introduce them into academic settings [23–25].

Margaret Dayhoff developed the first programs to determine the amino-acid sequence of a protein in 1965 and prepared the first protein sequence database that evolved into the PIR (protein information resource) in 1983. The sequence comparison and phylogenetic analysis were the first advances in the field of bioinformatics in the 1960s. Later, in the 1970s, structural analyses of macromolecules began. However, these analyses were quite limited due to the computing capacity available at that time. In that same period, computer methods also began to be applied in the processing of information about nucleic acids. Programs to compare sequences began to be developed. FASTA was developed around 1985, Genbank in the early 1980s, and SwissProt around 1987. In the late 1980s, the term bioinformatics started to be used for the science that integrated information technology and biology [26,27].

Even in the late 1980s, more advanced bioinformatics programs were developed in academic centers and quickly became commercial products, being distributed as integrated tool packages for the administration of molecular biology data. The improvement in computer systems allowed a great advance in automatic learning techniques with clear applicability in the field of bioinformatics. In the late 1990s, the demand for bioinformatics specialists was remarkable; however, few universities offered educational programs on this topic, which have grown considerably in recent years [23,24].

In 1999, Brazilian science stood out internationally with the complete DNA sequencing of the bacterium Xylella fastidiosa. This work relied on the use of genetic sequencing software based on the Internet, corresponding to the beginning of bioinformatics in Brazil. Currently, several sequencing projects are underway in our country, such as brGene, OMM, PIGS, Leifsoniaxyli, the genome of coffee, banana, and RioGene, among others [23,24].

In 2002, the first specialization course in bioinformatics was implemented in Brazil by the LNCC (National Laboratory for Scientific Computing),and in that same year, two strict sense postgraduate courses in bioinformatics were also authorized, one at USP and the other at UFMG, which are currently in full operation. Because of all that has been exposed, it's seen that bioinformatics is growing considerably, and it will be increasingly necessary for the interpretation of data in molecular biology. Sophisticated molecular techniques such as microarray, and new generation sequencing, among others, confirm the decisive role of bioinformatics in understanding the billions of data generated by these innovative tools [23,24].

Actually, in the world, bioinformatics has acquired increasing importance in the manipulation of biological data. Through the combination of procedures and techniques, it helps biologists with the complexity of both hardware and software tools, optimizing workflow in a distributed environment and reducing the overhead of data movement between programs. Thus, bioinformatics is a convergence of current technologies, which are in different stages of development, with the possibilities we still do not know what they will be. Since it is seen an extreme convergence between technological, physical, biological, social, cultural, and environmental means, and the transition from the digital revolution to a new industrial revolution, as a consequence of the computational influence, more and more accentuated and present, understanding the importance biotechnology and bioinformatics in this industrial evolutionary process, indicating an improvement in people's quality of life [3,28].

Bioinformatics is a knowledge to be applied in the field of biology. When it started to generate a large volume of sequencing data, for example, it was necessary to recruit computer scientists, statisticians, and mathematicians to develop software and tools to assist in these analyses. It seeks the integration of scientific knowledge with computational algorithms, for the generation of specific knowledge, contributing to new medical practices, facilitating the diagnosis, and indicating the best way to treat the individual patient [28,29].

Today there is research that generates data that cannot be analyzed without a computer: the Big Data analysis to reveal everything at the same time, not just one or two parts of a cell. Big Data is the term used to refer to the huge amount of structured and unstructured data generated every second. The keyword of bioinformatics is data integration, that is, huge amounts of data, developing methods that allow organizing and mining these vast amounts of data. Currently, a lot of biological, biochemical, and biophysical data have been produced in research, so the big idea that has to do with this is to be able to integrate and transform all this heterogeneous information into a result that can be interpreted and understood. The objective is to study and relate all this data to advance the research [30,31].

In the early 1970s, traditional biotechnology was a discipline confined to chemical engineering departments and microbiology programs. In 1971, the term "biotechnology" was coined by Hungarian engineer Karl Ereky. This researcher used the term to describe his experiment that aimed at the large-scale production of pigs fed with beets grown with microorganisms. Karl then defined biotechnology as all lines of research that involve the generation of products from raw materials that received the addition of living materials. However, this terminology remained quite ambiguous among scientists, and only in 1961, the term was then associated with the study of the industrial production of goods and services by procedures using biological organisms, systems, or processes. This is due to the Swedish microbiologist Carl GörenHedén, who suggested changing the title of a scientific publication journal in the field of applied microbiology and industrial fermentation titled *Journal of Microbiology and Biochemical Engineering and Technology* to *Biotechnology and Bioengineering* [32].

Traditional biotechnology focuses on three main aspects: (i) preparation of the raw material to be used as a source for microorganisms; (ii) the fermentation process of the material in bioreactors, obtaining the biotransformation and production of the desired material; and (iii) the purification of the final product. Obtaining a particular product on industrial scales is the main objective of biotechnology. So, a lot of research was done to improve the three aspects involved in the development of technology. For this purpose, several investments were made in the design of new bioreactors and the control and monitoring of fermentation processes. Despite a significant increase in production, the optimization of the biotransformation process remained below the desired level [33].

The strains of microorganisms capable of synthesizing the products of interest did so at suboptimal levels for an industrial scale. Random mutations induced by chemical mutagens and ultraviolet radiation were sometimes able to increase production levels. However, this scope is often limited. Usually, mutation induction affects not only the desired trait but other important ones for cellular metabolism [34].

Bioinformatics is an area of biotechnology that corresponds to the application of computational techniques to understand biological behavior in complex samples, considering advanced studies in cancer and chronic diseases using the tools available in computing, IoT devices, and new disruptive technologies, such as artificial intelligence and machine learning, in short cognitive computing. This results in more complex systems, complex research, transdisciplinary and multifunctionality, ranging from understanding gene alterations, as well as large-scale protein variations, impossible to be analyzed individually by manual techniques to the biological variability that exists between patients who are diagnosed with the same disease and can be measured and clarified with computer models, thus contributing to the development of targeted therapies, improving the prognosis of diseases that are difficult to diagnose and treat [35].

Bioinformatics is, therefore, an area of knowledge applicable in several institutions, such as clinical analysis laboratories, biotechnology, biochemistry or pharmaceutical companies, and also hospitals. It is an area of knowledge and not simply a platform for technological solutions. Its importance grows along with the increase in large-scale data generation. Finding various applications, including medicine, agronomy, and environmental sciences; development of new techniques and software for biology problems; and mathematical modeling based on networks, the development of programs that make it possible to study changes in molecule structures aims to produce more effective and more cost-effective drugs, which also finds multiple applications in biology and medicine [28,29].

These programs provide a consolidation in several areas of scientific knowledge, such as genetics, molecular biology, cell biology, microbiology, biochemical engineering, biochemistry, bioinformatics, biosafety, bioethics, among other diverse advanced segments. It has a multidisciplinary profile, which is in line with the guidelines of Translational Medicine, adding knowledge from the biosciences, informatics, and exact sciences for management, analysis, and even prognosis of medical and biological data, which are elements or measures collected from biological sources such as DNA, RNA, proteins and enzymes, digital images and other data [29].

Concerning multidisciplinary, it makes getting practical results faster, whether for clinical, surgical treatments, or technology development in the medical field. Just as also, the necessary structure for the application of translational medicine must meet this new concept of faster transfer of information from the laboratory to clinical practice. Companies must be aligned with academic research. These, in turn, can go from being merely theoretical to being applied, in a two-way street—the company understands what society wants, and the academy studies new technologies—because of the need to approach interdisciplinary areas to develop enabling solutions for the new industry based on engineering and its methods in modern health [3].

5.6 IoT IN HEALTHCARE

The IoT in medicine concerns the integration of medical devices into a communication network where information is exchanged and collected and the health sector has numerous possibilities for applicability, especially concerning the prevention of chronic diseases and reducing hospital infections, which are one of the main aggravating factors in the quality of life of patients and the high costs of institutions. Since the use of this technology in medicine has brought significant advances as a result of this application, more accurate diagnoses and improved quality of medical treatment have been achieved [21].

The automation and computerization of IoT processes in medicine bring not only more efficiency, when well applied, it can impact costs and improve the results of an organization, in addition to offering a more satisfying and humanized experience to the patient, but also more process security through intelligent devices that interact with each other to solve problems and generate efficiency, without the need for human help to assist in the process. Through IoT in medicine, it is possible to apply solutions that connect two devices where there will be an information transmission, which will allow greater autonomy of the patient and better monitoring in the treatment, considering the mobile applications that connect to other electronic devices to obtain data on the patient's health, sending accurate information that can assist the doctor in the detection, and treatment of possible diseases [36].

IoT in medicine is a technology that allows communication between equipment, improving medical care, and preventive diagnosis of diseases and surgeries helping to save lives. It is an important asset for the entire healthcare value chain, benefiting patients, hospitals, operators, research and development institutes, professionals, and pharmaceutical laboratories. One of the most obvious and popular applications of health services and IoT is remote health monitoring, known as telehealth,taking into account that in some cases, patients do not even need to visit an emergency room or hospital. This minimizes costs and eliminates the need for some visits,as well as helping to improve the patient's quality of life,saving them the inconvenience of travel,considering those patients with limited mobility or depending on public transport,something as simple as that can make a total difference [5].

One of the IoT applications in medicine, for example, is the monitoring of patients remotely using wearable devices and mobile applications. To quickly identify changes and reduce the incidence of serious illnesses, the data collected can, in realtime, feed electronic medical records in hospitals or medical clinics, assisting in decision-making. The benefits of applying IoT in medical environments can be seen as keeping patients more connected to doctors through remote monitoring and virtual visits, helping hospitals track employees and patients, automating the patient care workflow, maintaining quality control, and manage sensitive items while they are in transit, facilitating care and monitoring of chronic diseases, as well as lowering healthcare costs by simplifying the overall process [37].

5.6.1 IoT in Healthcare Applications

For those modern hospitals, the need for state-of-the-art software and hardware to operate is present, considering all those electronic devices, prone to countless risks ranging from power outages to system failures, can be a matter of life and death in certain cases of patients. And with the application of IoT devices in medicine instead of waiting for a device to fail, the system can take a proactive approach, virtually monitoring medical hardware and alerting hospital staff if there is a problem, processing alarms in realtime, allowing configuration performance rules and connections, and has low energy consumption [37].

As an efficient system in this sense, it is capable of continuously monitoring, in realtime, multiparameters such as temperature, door opening, consumption of medical gases, the pressure of the medical gas network, humidity, positive pressure, critical environment, and industrial machinery, among many other parameters; sending alerts when measurements are detected above or below the defined parameters; and sending automated compliance reports on scheduled time slices. What guarantees the protection of valuable assets is seen as IoT in healthcare [38].

In the same sense as the application of intelligent sensors, which is an IoT application in health, systems with IoT devices can remotely monitor the temperature remote control of vaccine refrigerators in clinics and health facilities, placed inside a refrigeration unit that sends data informing about the current temperature, providing access to real-time usage measurements so that public health professionals can more safely administer injections for diseases and save lives [39].

Just like barcode systems, or through a QR code that can be scanned with a smartphone camera, or RFID (radio frequency identification) tags that have batteries, active transmitters, and integrated electronics to capture and relay information and transmit data real-time medical diagnostics for doctors and other health professionals, which are health IoT applications, it enables public health situations, including infectious diseases, to be managed more effectively. In the same context of smart tags, an IoT application in health, through the application of IoT devices in medicine added to distribution mechanisms, allows physicians to maintain an accurate trajectory of medications and monitor whether patients are following their treatment plan correctly [40].

One of the main advantages of using IoT in medicine is the continuous monitoring of the patient, considering that there will be increased access to information on the patient's clinical condition, as well as more data to obtain a more accurate diagnosis, consequently, the possibility of offering personalized treatment. Since using IoT devices in medicine, it is possible to help in the treatment of chronic diseases through the combination of technology, data analysis, and mobile connectivity. Still because monitoring these data in realtime means saving time for health professionals and sharing information, such as diagnostic imaging exams, such as X-rays, since capable digital devices can be used of generating the exam data digitally, that is, IoT in Healthcare, is another significant benefit of using IoT in medicine [41].

Also, in the context of patient monitoring, considering the use of tools for mobile devices and the Web (health IoT applications), medical assistance, as well as through health professionals, is allowed to check treatment trends and respond more effectively to the data collected by these devices. The application of IoT technology in medicine can be done through a cloud platform connected wirelessly to various therapeutic devices in drugs, performing this activity in patients with chronic diseases, such as the case of the effectiveness of Parkinson's drugs and making necessary dosage adjustments in real time, where the collected data are transmitted to doctors by a wireless network, which, in turn, analyzes the patient's progress and the responsiveness to medication [39].

Since it is currently common to carry out an "episodic assessment" of patients with Parkinson's, that is, he goes to the clinic where he is subjected to tests that will assess the progress of the disease, even considering that the patient's performance will be different in comparison with your home environment and daily routine. And by performing this type of monitoring in the home environment, with the use of IoT in medicine, these patients will benefit from a more assertive treatment. The monitoring of a hospital bed through IoT sensors (health IoT applications) facilitates the continuous monitoring of vital signs data as well as the ability to monitor the patient's condition via cloud web platforms. With this, the doctor can perform a much more assertive treatment based on the patient's specific symptoms [42].

Another applicability of the technology are cardiac pacemakers that use IoT technology, where it is implanted in the patient, providing information at all times on the patient's cardiovascular system (health IoT applications),monitoring the patient's condition in realtime, and with that, any change or risk to their health is minimized by using much faster and more precise intervention. In modern home environments, IoT can be applied to a mattress with an application that detects the environment and sleep (health IoT applications), analyzing the data collected concerning breathing, heart rate, snoring, sleep environment, and temperature and sending the information directly to a smartphone or computer to improve the quality of ideal sleep. This data help to determine the best course of action for a better rest experience [36,38].

Still analyzing the application of IoT for health in the internal environment of homes, the technology may have resources aimed at helping babies sleep more and safely, including "cradle swing" and a cry sensor that automatically adjusts sound and movement (health IoT applications),preventing it from rolling, providing better quality sleep for babies, it also improves the quality of sleep for parents who may suffer from zinc deficiency due to sleepless nights. In addition to collecting information featuring a daily sleep record, it provides alerts on mobile devices and different settings to adjust according to the baby's age and sensitivity [36,38].

In a medication control context, IoT can be applied through a platform for doctors and healthcare professionals to maintain communication with patients and ensure that they keep track of medication dosage and frequency (health IoT applications),which in many cases through the use of sensors in a smart pill bottle with the medication that is used and issues reminders, which can be through text or smartphone of missed doses, or even to ensure that patients remember to take their prescriptions. This facilitates personalized support for medication refills, which eliminates the time and travel expenses, especially in distant locations, and reduces hospital costs and health problems as well as bringing real help to patients. IoT technology in medicine facilitates frequent monitoring of patients through monitored data, offering more specific treatments and transparency of compliance. This technology helps chronic patients to have a better quality of life, avoiding forgetting the doses of medication and reducing recurrent hospitalization due to the disease [37].

Analyzing a context of the application of IoT in the human body, an IEM (ingestible event marker) sensor device can be used which is an ingestible sensor the size of a sand grain, ingesting the sensor along with prescribed medication, generally used together with pills containing microscopic sensors that can send a signal to an external device, usually an adhesive used on the body, to ensure proper dosage and use, and with permission, a way for others to also see the collected information, which is activated in digestion, that is, a health IoT application [40].

In the technological context, using an IoT device in medicine it is possible to measure blood sugar levels through the tear fluid. Another health IoT application is the device consisting of a flexible metal coil covered with a layer of the hydrogel. Inside the hydrogel, there is an enzyme called glucose oxidase, which is the same used in blood sugar tests. And in the presence of glucose, this enzyme produces an electrical signal that is picked up by a nanosensor in the metal coil. Acting as a glucose monitor, it promises to accurately measure the sugar levels of patients with diabetes [43].

The technological scenario of IoT in health concerning pacemaker, IEM sensor, or the device applied to contact lenses is considered as digital medicine, which brings a new digital therapy that will advance the understanding of how patients are taking their medicines. This information along with the information that a dose of a certain digital medicine has been taken will integrate to give a broader view of how patients are functioning in the real world [44].

The IoT in the context of medicine can perform bed monitoring in hospitals, through a shared and unified system with IoT sensors and devices through the hospital, process bed requests at a time, and perform other parameters such as tracking the proximity of the nursing team. This technological approach to connecting and tracking hospital beds results in prior and accurate knowledge when a bed has been released and where it is located, reducing waiting times in emergency rooms or even patients in the hospital's emergency room. This type of technology can be useful in pandemic cases, such as those currently plaguing the world,COVID-19 [44].

In the context of hygiene in hospital environments, IoT can be used as sensor systems (health IoT application) that monitor and control hygiene practices in these environments, considering the proper frequency and following the correct hand washing standards, in addition to reporting in realtime compliance data to the correct standards. Considering sensors that can determine whether a member of the medical team washed their hands before and after interacting with the patient, the system records incidents of noncompliance which is considered an issue that needs to be addressed rigorously, preventing possible contamination and even the death of patients from infection, as in the case of sepsis. Similarly, through the installation of gel alcohol dispensaries connected by RFID tags (health IoT application) we can monitor the recurrence with which each professional performs the cleaning. This hygiene monitoring through IoT contributes to the preservation of the health of hospitalized patients by reducing the number of nosocomial infections by encouraging the hand hygiene of the institution's professionals, and it can also be useful in pandemic cases, such as COVID-19 [38].

Another application of IoT technology in hospital environments aims at safer surgeries through identification using chips and sensors in each of the instruments used in the operating room. It facilitates the traceability of prostheses, special surgical materials, devices, and materials of high added value, enabling the identification of the exact location of each instrument, preventing, for example, that one of them is forgotten inside the patient. The IoT in the context of medicine can also monitor the performance of MRI, as long as the device is equipped with a sensor that measures environmental factors in relation to certain limits by triggering alerts; in case the limits are exceeded, an alert can be sent by e-mail and text or even integrated into a local alarm system [39].

The use of IoT technology in health can also be applied in collecting information about the lifestyle of patients who are undergoing cancer cure treatment through the use of an activity tracker helping to record exhaustion and activity level. The information is collected daily through apps or wearables (health IoT applications), consisting of a useful technique that identifies and adjusts the treatment according to the need. Since the patient's reactions to treatment are necessary for personalized adjustments, it is usually done 1 week before treatment and then for several months during the treatment period [38].

Through IoT applied to healthcare, it is possible to note that patients can be connected to medical service providers, and they provide services remotely. Through the devices, it is possible to collect and send biomedical data and medical information collected through sensor readings, and other methods can be used to quickly analyze health problems, diagnose conditions, and provide real-time support. Through analysis it is possible to obtain several vital statistics of a patient, such as blood pressure and precision integrity scale, or even act as a personalized cardiac monitor, processing the sensor readings and visualizing the data through medical pulsation graphs [42].

From a medical care perspective, the ability to collect hospital patient information in realtime through a variety of IoT devices and sensors can ensure that healthcare professionals are notified immediately in an emergency and help doctors monitor the patients most effectively, which can mean the difference between life and death not only for patients in the ICU (Intensive Care Unit) or pregnant women who require continuous attention but for patients in serious conditions in general; obtaining continuous information about the status of a patient helps empower healthcare professionals to provide personalized care and respond when patients need them most [34].

IoT applications in the context of medicine allow internal connections between other devices and sensors, such as Bluetooth, Wi-Fi, and other connections, facilitating connectivity with a variety of medical devices. These examples of IoT in healthcare show how technology can be applied to countless resources for different purposes [36].

All of these IoT applications in healthcare are very promising, since it is the next frontier already present in patient care, both inside and outside the hospital, considering that the medical sector and related technological progress have been one of the slowest in adopting these new technologies. What is seen in the adoption of Big Data, IoT, and Artificial Intelligence, which have great potential in the healthcare sector, allows analyzing the data collected from connected devices and correlating them with information from the medical literature, generating insights for the entire value chain. For example, it is possible to identify certain regions of a country where there is a higher incidence of a certain pathology [36].

In this sense, using IoT devices streamlines medical and hospital processes that use IoT solutions to increase efficiency, through implementation in various institutions in the sector, especially in areas focused on patient care. IoT revolutionizes medicine and people's lives, with a proven return on investment in hospitals and laboratories, in addition to helping to reduce the rate of nosocomial infections. From the patient's point of view, it will be possible to access the entire medical history at any time and from any place based on information coming from connected devices, that is, the medical record of a lifetime will be concentrated, allowing a much more intelligent analysis, which is assertive and individualized [45–48].

5.7 DISCUSSION AND RECENT TRENDS

Medicine is one of the fields that most benefits from IoT through mobile applications connected to different devices; it is possible to relate data on the health of patients, including diagnostic and treatment data, and generate significant medical insights.

This connectivity also allows clinics and hospitals to monitor their users in real-time. With all this wealth of information enabling constant monitoring of patients and generating advances in diagnosis and treatment, diagnoses become more accurate and treatments become more efficient, which improves the quality of life of patients, which is aligned with the guidelines of translational medicine, and improves the welfare and health of society as the result of the process.

IoT in the context of medicine is the system by which semiautonomous devices and applications interact with each other to solve problems, generate opportunities, and ensure efficiency. From notification for nurses to update a patient's report card to medical decisions or guidance for the patient before or after the medical consultation, where both staff, doctors, and patients interact with new IoT technologies, observing clinical practice and changes in patient's necessary services, everything encompasses IoT in the context of medicine.

Privacy and security are also major concerns, and one of the biggest challenges of IoT in medicine is data security since information about the health of patients is confidential and since there is doctor–patient confidentiality in the Code of Medical Ethics, both during transmission and storage and sharing. In an emergency-room context, for example, data such as blood type, health history, and age of the patient serve to guide to arrive at possible diagnoses and treatments.

The technology to protect IoT devices exists. However, for a device, object, or component connected to IoT to be secure, it must not have an open connection, that is, publish, but should be encrypted sufficiently to be able to guarantee the protection of the object. Thus, a VPN connected to a central routers in private networks, considering encryption features and access allowed only for authorized persons, enabling protection, as well as preventing loss and theft of valuable information.

Several devices can be connected to the Wi-Fi network or others in a hospital, generating large amounts of data and allowing the tracking of information. In the case of patients, they can be monitored using medical equipment that sends messages to the nursing staff or specialists on call when necessary, being useful in all departments of a hospital.

Bioinformatics is the science that researches, develops, and innovates working directly with the data generated by medical and biological research methodologies, where such data can come from IoT devices and components. Being an essentially multidisciplinary discipline, it makes intercession between several areas of activity, which uses computer science methods for the acquisition, management, analysis, and prediction of biological and medical information, and transforms this into knowledge.

Currently, the applications of bioinformatics consist of genome assembly, comparative genomics, gene expression analysis, gene regulation networks, the study of metabolism, analysis of the structure of macromolecules, drug design, and evolutionary biology. Combining the applicability of bioinformatics with biotechnology, we noted a great performance of both in pharmaceutical industries. To understand this fact, we need to consider that the proper functioning of drug therapy is the product of the interaction of several molecules, which when in contact with the biological target in the patient's organism inhibit or accelerate certain chemical reactions, aiming at the attenuation of the pathological condition. The formulation of a drug has numerous stages and tests, both preclinical and clinical, which aim to evaluate toxicity, efficacy, and side effects, among other aspects. For the production of new drugs, it is essential that biological molecules are selected with the necessary profile for the use of that drug and that other molecules are discarded, to not interfere with the selected molecules.

Based on this, joining bioinformatics, biotechnology, 5G networks, IoT, and wearable technologies may seem ambitious but very promising today. Through Smart Grids 5G networks, patients can be monitored 24h a day by their doctors, both in relation to the heartbeat, temperature, and glucose, and to the performance of periodic analyses of DNA structures to detect the presence of polymorphisms and gene mutations at an early stage without the patient showing clinical signs. In other words, the union of wearable devices, smart grids, and bioinformatics can generate extremely early diagnoses, increasing the rate of survival and cure for patients.

However, developing only IoT devices in the context of medicine and wearable technologies for early detection of diseases of genetic origin is not enough to ensure the well-being of all people, as many diseases are caused by external agents, such as viruses, fungi, parasites, and bacteria. In this way, investing in diagnostic methods for the detection of infectious diseases through wearable devices, which, using microliters of blood, performs a search for DNA from external agents makes medical diagnoses much more accurate.

Another scenario of high applicability of IoT in the context of medicine, wearable technologies, bioinformatics, and biotechnology is the research area, mainly those aimed at the sectors of development of innovative vaccines and pharmaceuticals. This possibility would be of great use especially amid pandemics, such as that caused by COVID-19. In addition, many experts claim that even in this century, humanity could be affected by another pandemic, equal or worse to that caused by the coronavirus. In this context, drug efficacy depends primarily on the result of interactions between the body's cells with the developed molecules. In this way, any type of wearable technology, such as watches or bracelets, could constantly analyze the patient's blood to seek out and describe drug interactions in realtime. Such a reality would avoid invasive procedures and frequent trips to hospitals or research centers; after all in times of pandemic, the first measure to be adopted is social isolation. Thus, with remote monitoring, it is possible to reduce the rate of deaths and incidence of cases in healthcare professionals.

One of the most popular applications of health services and IoT is remote health monitoring that minimizes costs and eliminates the need for some visits to hospitals and medical centers, helping to improve the patient's quality of life, saving them the inconvenience of the trip. Provided that if a given patient has limited mobility or depends on public transport, this technological possibility makes a difference, as intelligent technology helps to simplify care for both patients and their caregivers.

The sharing of data captured by digital devices during diagnostic tests, carried out via the telemedicine platform, is an example. This information is automatically added to the patient's record and is available for interpretation by specialists, enabling the issuance of distance reports with quality and efficiency. The data shared through the telemedicine platform are accessed by qualified specialists, responsible for interpreting the records. And with the support of the information available in the patient's record, the specialists evaluate the exams and produce the digitally signed report at a distance. So, technology makes interaction with the doctor much more powerful and useful, as well as giving more control.

As the volume of this data is enormous, especially when we use new-generation sequencing equipment, it is necessary to process, integrate, and analyze all the information, making it possible to also compare the data generated in the laboratory and the various existing databases and making the analysis and release of the patient's result faster. Although from the computational point of view of bioinformatics, the medical/biological data can be treated in the same way as any other, the results and conclusions of its analysis are strongly dependent on the experimental methodology by which they were obtained. Since computational, statistical, and mathematical methods are necessary to organize and process this data, they must also be analyzed through biological lens, taking into account the particularities of living systems. This is necessary to choose the best approach for data analysis and to facilitate dialogue with the multidisciplinary involved through translational guidelines applicable to the biological and health areas.

However, ensuring your protection on the Internet requires a series of measures, starting with setting limits, which can be done through regulation. On the other hand, the variety of standards created by different countries to govern the use of health data can hinder the globalization of innovations. The development and integration of systems is another barrier that prevents the popularization of IoT in health since different devices must be able to exchange information with each other. Through modern and intuitive platforms, registering and storing information transmitted by connected devices, taking advantage of the potential of IoT to expand the medical technological portfolio, and reducing costs is made possible.

Digital devices for diagnostic tests, such as X-rays, tomography, and electrocardiogram, among many others, are capable of generating and storing data digitally, eliminating the use of paper. This possibility preserves the environment, reduces the need for physical space for files, and guarantees the conservation of documents, without any additional care as in the handling of radiographic films, which could be easily damaged. Digitally archived, the information allows for greater simplicity in sharing, which can be done via email or messaging applications.

From the infrastructure point of view, when redesigning hospital processes to accommodate the technology, several unclear workflow inefficiencies are identified and corrected, where the effect caused by IoT in the context of medicine and how much is the reformulation of patient care are positive.

From the patient's point of view, some connected devices can be implanted in it, for example, smart pacemakers, which facilitate the monitoring of thousands of people around the world. Like wearables, these devices collect, store, and send information in realtime about the patient's cardiovascular system. Equipped with applications and other wearable systems, they can monitor symptoms and behaviors of people with depression, such as mood and cognition assessments, while daily assessments made from questionnaires and other data corresponding to test results are not as efficient.

Whether to facilitate the work of medical and scientific research teams, making processes more efficient, or to improve the patient's living conditions, IoT in health is a discovery that brings benefits that are reflected in the success of treatments, in reducing losses and damages, and in improving hospital management, impacting people's lives in a very positive way. In this way, the era of the machine and smart devices replaces traditional IT people and creates entirely new fields. This scenario is similar to what happens with traditional medicine, related to adapting or perishing.

Two distinct areas related through bioinformatics in relation to the large database useful for carrying out studies on the diseases and side effects of medicines, for example, allowing execution in almost realtime. The focus is not on the use of IoT itself in translational medicine, but the data extracted from their own information systems.

5.7.1 Recent Trends

The trend is that the presence of objects connected to the Internet is increasing in the health area, the number of gadgets capable of capturing and helping to monitor signs and health data, diseases and characteristics of each individual, appear several wearable devices, including new health features, such as an accelerometer and a gyroscope, capable of detecting sudden drops, still containing electrodes and an electrical heart rate sensor that can take an electrocardiogram, thus proposing an even greater integration of the human body with technology [2].

The trend in the home healthcare market with IoT applications and devices geared toward "health tech" allows for videoconferencing and wearable devices. These connected devices, internal sensors, and the collected data allow individuals to maintain the lives of independent patients at a much lower risk. The crossing of information from different sources and technologies, which talk to each other, helps to speed up diagnoses, thus providing patients with the most relevant information about their health. Hospitals and healthcare professionals use IoT to keep patients remotely connected to healthcare providers and services, controlling and monitoring the patient's vital signs and health status indicators through health devices [37, 46–48].

No matter what the shape, the fact is that all of these devices or sensors together will further revolutionize medicine and patients' lives. Since analogous to how sight, hearing, smell, touch, and taste allow people to understand the world, IoT sensors in the context of medicine allow machines to understand the world. In the future, it will be natural and possible to access health data and the entire patient history anytime, anywhere, based on information from connected devices. In other words, all the information about each patient, such as allergies, medical preferences, chronic diseases, and other relevant data, are stored safely in the same place, significantly reducing visits to medical offices and hospitals and decreasing general health costs [4–6,46–48].

The trends related to bioinformatics are in the use of databases (genomic, molecular, and derivative data), so that soon, through the translational guidelines, the integration of several research institutes that work with genomic and molecular data can obtain discoveries about the genome be made more quickly and with greater efficiency [10].Due to its multidisciplinary nature, which is in line with translational guidelines, biotechnology covers various segments such as human health, industry, and bioinformatics, among others, which have been allowing new applications such as processes involving genetic engineering. This technique allows the manipulation and transfer of genetic material. Since bioinformatics assists in the development of these processes, it maps and organizes the biomolecular sequences of the various living organisms, allowing them to be manipulated [10,12].

5.8 CONCLUSION

Bioinformatics is a sum of two areas of knowledge: computational science applied to Biomedicine, as it is a multidisciplinary area with double vision, it needs to navigate in these two areas, since genes, genomes, molecular modeling, metabolic systems, and also biological models are structures with so much information that only through information technology or computer science has properties for to process, organize and transmit this data. Through Bioinformatics, it is also qualified to work with image and signal processing, which aid medical diagnosis, and management of clinical and public health records, which are systems that streamline information and improve medical care.

The IoT in the context of medicine consists of the digital communication between physical objects (devices), computing, data collection, and transmission, and device control properties and characteristics, that is, through sensors or actuators that allow control, i.e., the digital health control system. These "things" must have these characteristics and should be all-digital, which in the sense of IoT in medicine are complete digital objects, used in the daily medical routine, using information from sensors at all levels, from hospital treatment to the monitoring of the population's health conditions, exchanging data and information among themselves and with their users over the Internet. Medical IoT technologies can automatically gather the necessary information from patients and detect possible diseases to prevent them. It aims to facilitate by promoting greater productivity and reducing costs, resulting in both the patient and the doctor a little more health control more technologically and simply, as this technology is actively integrated with different medical devices.

In the field of bioinformatics, as well as in translational medicine, they are recognized for their significant contributions to the areas of carcinogens, epidemiology, molecular and cellular biology, tumor microenvironment, and new therapeutic strategies. Through technology it is possible to accelerate the interaction and collaboration between researchers from different areas, contributing to the improvement of translational studies. Through the use of analytical methods and modern bioinformatics tools to better understand the mechanisms of action and/or the signaling pathways of a disease, it is possible to identify potential molecular biomarkers and/or promote scientific support for the development of early diagnostic methods of a disease and/or the development of biotechnological products with potential use in public health.

REFERENCES

1. Krawiec, P., Sosnowski, M., Batalla, J. M., Mavromoustakis, C. X., Mastorakis, G., & Pallis, E. (2017). Survey on Technologies for Enabling Real-Time Communication in the Web of Things. In *Beyond the Internet of Things*, J. M. Batalla, G. Mastorakis, C. Mavromoustakis, & E. Pallis (Eds.), (pp. 323–339). Switzerland: Springer.
2. Yuehong, Y. I. N., Zeng, Y., Chen, X., & Fan, Y. (2016). The internet of things in healthcare: An overview. *Journal of Industrial Information Integration*, 1, 3–13.
3. Baxevanis, A. D., Bader, G. D., & Wishart, D. S. (Eds.). (2020). *Bioinformatics*. Hoboken, NJ: John Wiley & Sons.
4. Patade, A. S., Gandhi, H. P., & Sharma, N. (2019, January). IoT Solutions for Hospitals. In *2019 11th International Conference on Communication Systems & Networks (COMSNETS)* (pp. 813–816). New York: IEEE.
5. Chatterjee, P., Cymberknop, L. J., & Armentano, R. L. (2018). IoT-based eHealth toward decision support system for CBRNE events. In *Enhancing CBRNE Safety & Security: Proceedings of the SICC 2017 Conference* (pp. 183–188).Cham: Springer.
6. Bodur, G., Gumus, S., & Gursoy, N. G. (2019). Perceptions of Turkish health professional students toward the effects of the internet of things (IoT) technology in the future. *Nurse Education Today*, 79, 98–104.
7. Li, D., Azoulay, P., & Sampat, B. N. (2017). The applied value of public investments in biomedical research. *Science*, 356(6333), 78–81.
8. Pettibone, K. G., Balshaw, D. M., Dilworth, C., Drew, C. H., Hall, J. E., Heacock, M.,... & Walker, N. J. (2018). Expanding the concept of translational research: making a place for environmental health sciences. *Environmental Health Perspectives*, 126(7), 074501.
9. Filkins, B. L., Kim, J. Y., Roberts, B., Armstrong, W., Miller, M. A., Hultner, M. L., ... & Steinhubl, S. R. (2016). Privacy and security in the era of digital health: what should translational researchers know and do about it? *American Journal of Translational Research*, 8(3), 1560.
10. Guo, X., & Liu, J. (2018). Basic Concept of Translational Medicine. In *Atlantoaxial Fixation Techniques*, Bin Ni, Xiang Guo, & Qunfeng Guo (Eds.), (pp. 33–36).Singapore: Springer.
11. Fort, D. G., Herr, T. M., Shaw, P. L., Gutzman, K. E., & Starren, J. B. (2017). Mapping the evolving definitions of translational research. *Journal of Clinical and Translational Science*, 1(1), 60–66.
12. Lamb, J. A., & Curtin, J. A. (2019). Translational medicine: insights from interdisciplinary graduate research training. *Trends in Biotechnology*, 37(3), 227–230.
13. Gupta, V., Sengupta, M., Prakash, J., & Tripathy, B. C. (2017). An Introduction to Biotechnology. In *Basic and Applied Aspects of Biotechnology* V. Gupta, M. Sengupta, J. Prakash, & B. C. Tripathy (Eds.), (pp. 1–21). Singapore: Springer.
14. Morrison, B. W. (2016). Biotechnology and translational medicine. *Clinical and Translational Science*, 9(3), 125.
15. Plenge, R. M. (2016). Disciplined approach to drug discovery and early development. *Science Translational Medicine*, 8(349), 349ps15–349ps15.
16. França, R. P., Iano, Y., Monteiro, A. C. B., & Arthur, R. (2020). Intelligent Applications of WSN in the World: A Technological and Literary Background. In *Handbook of Wireless Sensor Networks: Issues and Challenges in Current Scenario's*, P. K. Singh, B. K. Bhargava, M. Paprzycki, N. C. Kaushal & W.-C. Hong (Eds.), (pp. 13–34). Cham:Springer.
17. Balani, N., & Hathi, R. (2016). *Enterprise IoT: A Definitive Handbook*. CreateSpace IndependentPublishing Platform. Scotts Valley, CA: Amazon.

18. COSTA, Felipe Camargos. Análise de modelos de representação de informação normativa usando XML. 2017. viii, 51 f., Il. CourseConclusionPaper (Bachelorof Computer Science) —Universidade de Brasília, Brasília, 2017
19. Miyanaga, Y., Watanabe, H., & Osada, N. (2017). *Big-Data Analysis, IoT and Bioinformatics.*
20. Monteiro, A. C. B., Iano, Y., & França, R. P. (2017). An improved and fast methodology for automatic detecting and counting of red and white blood cells using watershed transform. *VIII Simpósio de Instrumentação e Imagens Médicas (SIIM)/VII Simpósio de Processamento de Sinais da UNICAMP.*
21. França, R. P., Iano, Y., Monteiro, A. C. B., Arthur, R., Estrela, V. V., Assumpção, S. L. D. L., & Razmjooy, N. (2019). Potential Proposal to Improvement of the Data Transmission in Healthcare Systems.In *VI InternationalSymposiumonImmunobiologicalsand VII Seminário Anual Científico e Tecnológico.* Bio-Manguinhos/Fiocruz.
22. Monteiro, A. C. B., Iano, Y., França, R. P., & Arthur, R. (2020). Development of a Laboratory Medical Algorithm for Simultaneous Detectionand Counting of Erythrocytes and Leukocytes in Digital Images of a Blood Smear. In *Deep Learning Techniques for Biomedical and Health Informatics,* A. Abraham, A. Kelemen, M. Mittal, S. Dash, & B. R. Acharya (Eds.), (pp. 165–186). Cambridge, MA: Academic Press, Elsevier.
23. Sansom, C. (2006). The beginnings of bioinformatics. *The Biochemist*, 28(6), 48–49.
24. Gauthier, J., Vincent, A. T., Charette, S. J., & Derome, N. (2019). A brief history of bioinformatics. *Briefings in Bioinformatics*, 20(6), 1981–1996.
25. Daniel, B. (2019). *Nanopore Sequencing: An Introduction.* Singapure: World Scientific.
26. Henzy, J. E. (2018). *Margaret Dayhoff: Catalyst of a Quiet Revolution. Women in Microbiology*, 29–36. Hoboken, NJ: Wiley.
27. McGarvey, P. B., Huang, H., Barker, W. C., Orcutt, B. C., Garavelli, J. S., Srinivasarao, G. Y.,… & Wu, C. H. (2000). PIR: a new resource for bioinformatics. *Bioinformatics*, 16(3), 290–291.
28. Hobbs, B. P., Berry, D. A., & Coombes, K. R. (2020). Biostatistics and Bioinformatics in Clinical Trials. In*Abeloff's Clinical Oncology*, J. Niederhuber, J. Armitage, J. Doroshow, M. Kastan, & J. Tepper (Eds.), (pp. 284–295, 6th Edition). Cambridge, MA: Elsevier.
29. Vanderbilt, C., & Middha, S. (2020). Role of Bioinformatics in Molecular Medicine. In *Genomic Medicine* (pp. 55–68). Cham: Springer.
30. França, R. P., Iano, Y., Monteiro, A. C. B., & Arthur, R. (2020). Improved Transmission of Data and Information in Intrusion Detection Environments Using the CBEDE Methodology. In *Handbook of Research on Intrusion Detection Systems*, S. Srinivasagopalan & B. B. Gupta (Eds.), (pp. 26–46). Pensilvânia, PA: IGI Global.
31. França, R. P., Iano, Y., Monteiro, A. C. B., & Arthur, R. (2020). Lower Memory Consumption for Data Transmission in Smart Cloud Environments with CBEDE Methodology. In *Smart Systems Design, Applications, and Challenges*, P. J. S. Cardoso, J. Monteiro, J. Rodrigues, & C. M. Q. Ramos (Eds.), (pp. 216–237). Pensilvânia, PA: IGI Global.
32. Fári, M. G., & Kralovánszky, U. P. (2006). The founding father of biotechnology: Károly (Karl) Ereky. *International Journal of Horticultural Science*, 12(1), 9–12.
33. Khan, F. A. (2011). *Biotechnology Fundamentals.* Boca Raton, FL: CRC Press, Taylor & Francis Group.
34. Oztemel, E., & Gursev, S. (2020). Literature review of Industry 4.0 and related technologies. *Journal of Intelligent Manufacturing*, 31(1), 127–182.
35. Ali, M. A., Alexiou, A., & Ashraf, G. M. (2019). Biotechnology and Bioinformatics Applications in Alzheimer's Disease. In *Biological, Diagnostic and Therapeutic Advances in Alzheimer's Disease*, G. M. Ashraf & A. Alexiou (Eds.), (pp. 223–234). Singapore: Springer.

36. França, R. P., Iano, Y., Monteiro, A. C. B., & Arthur, R. (2020). Potential proposal to improve data transmission in healthcare systems. In *Deep Learning Techniques for Biomedical and Health Informatics*, A. Abraham, A. Kelemen, M. Mittal, S. Dash, & B. R. Acharya (Eds.), (pp. 267–283). Cambridge, MA: Academic Press, Elsevier.
37. Farahani, B., Firouzi, F., & Chakrabarty, K. (2020). *Healthcare IoT. In Intelligent Internet of Things* (pp. 515–545). Cham: Springer.
38. Rajaseskaran, R., Jain, R., & Sruthi, M. (2020). Patient Health Monitoring System and Detection of Atrial Fibrillation, Fall, and Air Pollutants Using IoT Technologies. In *Incorporating the Internet of Things in Healthcare Applications and Wearable Devices* (pp. 165–183). IGI Global.
39. Kaur, H., Atif, M., & Chauhan, R. (2020). An Internet of Healthcare Things (IoHT)-Based Healthcare Monitoring System. In *Advances in Intelligent Computing and Communication* (pp. 475–482). Singapore: Springer.
40. Malarvizhi, N., Annamalai, S., Raj, P., Neeba, E. A., & Lin, J. W. (2020). Industrial IoT Application Architectures and Use Cases. Boca Raton, FL: CRC Press, Taylor & Francis Group.
41. Dalal, P., Aggarwal, G., & Tejasvee, S. (2020). Internet of Things (IoT) in Healthcare System: IA3 (Idea, Architecture, Advantages and Applications). *Proceedings of the International Conference on Innovative Computing & Communications (ICICC) 2020.*
42. Sadoughi, F., Behmanesh, A., & Sayfouri, N. (2020). Internet of Things in Medicine: A systematic mapping study. *Journal of Biomedical Informatics*, 103, 103383.
43. Sujaritha, M., Sujatha, R., Nithya, R. A., Nandhini, A. S., & Harsha, N. (2020). An Automatic Diabetes Risk Assessment System Using IoT Cloud Platform. In *EAI International Conference on Big Data Innovation for Sustainable Cognitive Computing* (pp. 323–327). Cham: Springer.
44. Abbas, S. (2020, April). Cyber-Medicine Service for Medical Diagnosis Based on IoT and Cloud Infrastructure. In *Proceedings of the International Conference on Artificial Intelligence and Computer Vision (AICV2020)*, A.-E. Hassanien, A. T. Azar, T. Gaber, D. Oliva, & F. M. Tolba (Eds.), (pp. 617–627). Cham: Springer.
45. Akkaş, M. A., Sokullu, R., & Çetin, H. E. (2020). Healthcare and patient monitoring using IoT. *Internet of Things*, 11, 100173.
46. Raza, K. &Qazi, S. (2019). *Nanopore Sequencing Technology and Internet of Living Things: A Big Hope for U-Healthcare. Sensors for Health Monitoring* (Vol. 5, pp. 95–116). Amsterdam, Netherlands: Elsevier. https://doi.org/10.1016/B978-0-12-819361-7.00005-1
47. Qazi, S. & Raza, K. (2020). *Smart Biosensors for an efficient Point of Care (PoC) Health Management. Smart Biosensors in Medical Care* (pp. 65–85). Amsterdam, Netherlands: Elsevier. https://doi.org/10.1016/B978–0-12-820781–9.00004-8
48. Raza, K. (2014). Clustering analysis of cancerous microarray data. *Journal of Chemical and Pharmaceutical Research*, 6(9): 488–493.

6 Viroinformatics and Viral Diseases: A New Era of Interdisciplinary Science for a Thorough Apprehension of Virology

Kayenat Sheikh and Khalid Raza
Jamia Millia Islamia

CONTENTS

6.1 INTRODUCTION

It is evident that viruses have been causing infection since the 15th BC and have been fatal for centuries. It is therefore significant to learn about them with perceptiveness. They are found to exist in all life forms like bacteria, fungi, plants, and animals. Pathogenic viruses in humans cause deadly diseases such as HIV AIDS, influenza, SARS 2003, MERS, yellow fever, Ebola, Zika, swine flu, and the most recent COVID-19 (Sharma et al., 2015; Hoffmann et al., 2020a; Raza, 2020; Raza et al., 2020). Despite the development of vaccines and therapeutic drugs, and several treatments available, viral diseases keep on emerging and reemerging from time to time. Approximately 6 million people die each year globally because of viruses.

Therefore, it is crucial to develop remedies against these pathogenic invaders (Sharma et al., 2015). The risk of viral emergencies in recent years has increased exponentially due to several factors such as social, environmental, ecological, and economic. The changing climate, deforestation, urbanization, population explosion, and the movement of livestock, humans, and disease vectors are all responsible for causing pandemic-like situations. Taking into consideration the current global pandemic of COVID-19, studies suggest that it originated from the wet seafood market of Wuhan in China (Chakraborty and Maity, 2020). Pangolins to bats, every kind of wild livestock with strong host capabilities are consumed by people. Therefore, it should be regulated with health priorities. Urbanization and deforestation contribute to worsening pandemic situations. The more the green cover is cleared, the more the land gets converted into cities and towns (Chakraborty and Maity, 2020). As a result of this, the population explosion has occurred in many countries including China and India. The heavy population is the major source of community transmission, thus adding to pandemic situations. Climate change due to global warming, fluctuations in temperature, and atmosphere might favor viral growth and proliferation. The economic losses are enormous too (Bashar et al., 2018). The budget is calculated at 150 billion US dollars/year for all global disasters, out of which 30 billion dollars are spent on infectious disease outbreaks. Livestock damage is the worst of all affecting health and economy. It is also interesting to note here that all viruses are not pathogenic. Some are known to store information about their host and begin to influence the entire biogeochemical cycles (Bashar et al., 2018).

Recently, the advancements in the fields of molecular biology, computer science, and bioinformatics have resulted in the accumulation of data in bulk about the genes, genomes, and experiments. For storage, examination, and analysis from all these useful information and data, more than 150 viroinformatics resources have been developed to date (Sharma et al., 2015). We will learn about some of the tools and web resources in the upcoming sections. This chapter is divided into two distinct halves. In the first half, we will learn about the viruses, their taxonomy, evolution, and mechanism of action, effects caused by them, drugs, vaccines, and therapeutics discovery. In the second half, we will learn about the viroinformatics tools and resources that can be used to learn virology.

6.2 VIROLOGY: AN OVERVIEW

6.2.1 Brief History

The story began when the smallpox virus turned out to be the greatest killer of humanity across the globe. Edward Jenner's break through viral treatment in 1796 changed the history of viruses forever. He inoculated patients with cowpox lesions and it protected them from more lethal viruses. This led to the beginning of the new virology study phase. The only disease to be completely eradicated from the world is smallpox and this happened with the vaccination efforts of WHO in the year 1979 (Enquist et al., 2009).

The phase 1796–1885 has been classified as the "Protovirology" era. In this era, viruses were not even clearly identified. However, Jenner's discovery in 1796, the transmission of tobacco mosaic virus (TMV) disease with cell-free extracts, and the development of the rabies vaccine in 1885 were the path-breaking scientific discoveries (Koprowski, 1993).The next era is called "Auroravirology" or "dawn of virology." Between the years 1892 and1933, the rigorous study of viruses took place (Figure 6.1). Ivanovsky described TMV in 1892; development of the virus as a contagious element came into the picture through Beijerinck followed by discoveries of the first animal virus, first human virus (yellow fever), rabies virus, leukemia virus, poliovirus, RSV (solid tumor virus), measles virus, human influenza virus, etc. (Enquist et al., 2009).The next era is called as "Meridiovirology" that occurred in 1934–1955. In this era, the viruses affecting bacteria also called bacteriophages were briefly described. It was discerned that bacteriophage is composed of nucleic acid and proteins. The crystallization of the TMV and the in vitro assembly of infectious TMV from purified RNA and protein were done for the first time (Koprowski, 1993). The next era "Janovirology" 1956–1975 spans classical virology and the beginning of modern virology (sequence information).

In this era, gene expression, regulation, and development of important tools and techniques for cloning and restriction sequence mapping were carried out. The last and most recent is the "Neovirology." It marks the phase where the viral genomes were completely sequenced. The structures of the virus up to the atomic resolution were seen. This era began back in 1976 and continues to be present until today (Koprowski, 1993; Enquist et al., 2009).

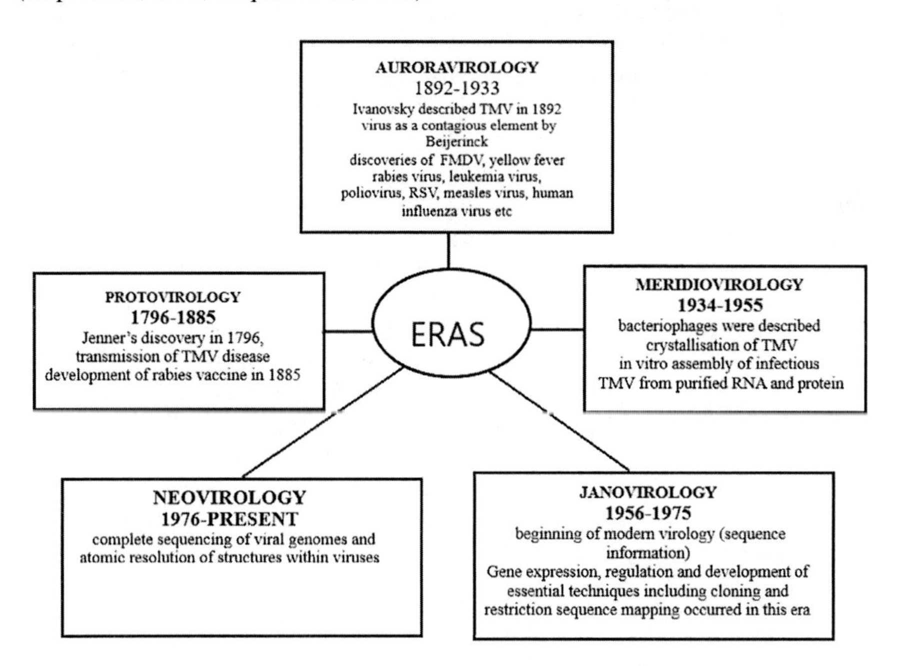

FIGURE 6.1 The eras of virology and path-breaking achievements (see Koprowski, 1993).

6.2.2 Structure, Function, and Classification of Viruses

Viruses are very tiny, obligate, and intracellular pathogens. They are not living organisms rather particles. They are fully dependent on the host cell for sustaining their pathogenicity. The host cell provides very complex metabolic and biosynthetic machinery. The viral genome is very unstable and hence can mutate at a very fast rate. The genetic material can be DNA or RNA, and it can be either single-stranded or double-stranded (Table 6.1) (Gelderblom, 1996; Simmonds et al., 2018).The protective protein layer called capsid protects the genome. The viral genome and some basic proteins are packaged inside the protein capsid. The nucleic acid-associated protein is called nucleoprotein. Capsid and nucleoprotein together form the nucleocapsid. In the case of enveloped viruses, the nucleocapsid is walled with a lipid bilayer derived from the host cell membrane attached with a glycol-proteinaceous outer layer of the virus envelope. A virion is a complete virus particle (Gelderblom, 1996).Virion transfers its genetic material (DNA or RNA) inside the host cell. The genome is transcripted and translated by the duplication machinery of the host cell. Table 6.1 provides an insight into the various categories of viruses used to classify them.

There are in all seven categories formed by the virologists which include Group I (dsDNA), Group II (ssDNA), Group III (dsRNA), Group IV (+ssRNA), Group V (–ssRNA), Group VI, and Group VII of RT viruses. Group I comprises all the dsDNA viruses, *Herpesvirales* viruses fall into this group. *Adenoviridae*, *Papillomaviridae*, and *Polydnaviridae* have been known to be pathogenic. Group II includes ssDNA viruses such as *Germiniviridae*, *Nanoviridae*, *Parvoviridae*, *Bodnaviridae*, and many more pathogenically active viruses. Group III consists of all the dsRNA viruses such as *Endomaviridae*, *Chrysoviridae*, and *Reoviridae*. Similarly, Group IV includes positive sense (+) ssRNA viruses. Most viruses belong to this group. *Nidovirales*, *Tymovirales*, and *Picornavirales* are some of the examples. Group V has all the negative sense (–) ssRNA genomic viruses, for example, *Mononegavirales*, *Jingchuvirales*, *Serpentovirales*, *Goujianvirales*, *Muvirales*, *Anticulavirales*, and *Bunyavirales*. Groups VI and VII include all the RT viruses such as *Ortevirales*, *Caulimoviridae*, and *Hepadnaviridae*. By now, a question might have arisen in your mind, whether the genome types DNA/RNA influences the pathogenicity of any virus. Or which type of virus is more virulent? Well, it is quite unclear to call either of them virulent. The RNA viruses are more in number compared to DNA viruses. The genome of RNA viruses is comparatively smaller. It mutates at a faster rate and very few proteins are synthesized by it (Durmuş and Ülgen, 2017). The DNA viruses on the other hand have longer sequences and integrate well with the host. They finely exploit the machinery of their host. Both DNA and RNA viruses have distinct infection strategies. Hence, it is too early to conclude either of them pathogenic. Rigorous studies and research on various pathogenicity parameters would probably answer to this (Durmuş and Ülgen, 2017)

6.2.3 Frequent Outbreaks of Viral Diseases

The emergence and reemergence of viral diseases in the past two decades have been causing a great threat to human lives. Viral diseases such as MERS, SARS,

TABLE 6.1
Classification of Viruses Based on Genetic Material (Simmonds et al., 2018)

Group I dsDNA	Group II ssDNA	Group III dsRNA	Group IV +ssRNA	Group V –ssDNA	Group VI and VII RT
Poxviridae	*Geminiviridae*	*Endomaviridae*	*Picornavirales*	*Mononegavirales*	*Ortevirales*
Iridoviridae	*Genomoviridae*	*Hypoviridae*	*Solinviviridae*	*Jingchuvirales*	*Retroviridae-Betaretrovirus*
Marseilleviridae	*Circoviridae*	*Totiviridae*	*Caliciviridae*	*Serpentovirales*	*Retroviridae-Alpharetrovirus*
Asfarviridae	*Smacoviridae*	*Botybirnaviridae*	*Potyviridae*	*Goujianvirales*	*Retroviridae-Lentivirus*
Phycodnaviridae	*Nanoviridae*	*Quadriviridae*	*Astroviridae*	*Muvirales*	*Retroviridae-Deltaretrovirus*
Mimiviridae	*Bacilladnaviridae*	*Chrysoviridae*	*Flaviviridae*	*Qinviridae*	*Retroviridae-Gammaretrovirus*
Bculoviridae	*Bidnaviridae*	*Magabirnaviridae*	*Carmotetranviridae*	*Articulavirales*	*Retroviridae-Epsilonretrovirus*
Nudiviridae	*Parvoviridae*	*Partitiviridae*	*Tombusviridae*	*Bunyavirales*	*Retroviridae-Spumavirus*
Herpesvirales	*Anelloviridae*	*Picobirnaviridae*	*Luteoviridae*		*Caulimoviridae*
Nimaviridae		*Amalgaviridae*	*Sobemoviridae*		*Hepadnaviridae*
Polydnaviridae		*Reoviridae*	*Tymovirales*		
Adenoviridae		*Birnaviridae*	*Virgaviridae*		
Lavidaviridae			*Bromoviridae*		
Papillomaviridae			*Closteroviridae*		
Polumaviridae			*Togaviridae*		
			Hepeviridae		
			Alphatetraviridae		
			Benyviridae		
			Rubiviridae		
			Nidovirales		
			Permutotetraviridae		
			Narnaviridae		

Marburg, Ebola, Lassa fever, Rift valley fever, Zika, Nipah and henipavirus disease, Chikungunya, and the most recent and deadly SARS-CoV-2 (Rothan and Byrareddy,2020) have created much havoc (Zumla and Zumla,2019).

Zoonotic coronaviruses of humans have a disastrous epidemic potential. They have emerged frequently in the past two decades. The severe acquired respiratory syndrome (SARS +ssRNA, a novel b coronavirus) emerged first in November 2002 in China and then spread worldwide at an alarming rate. Vaccines for SARS-CoV were tested on animal models, and inactivated whole virus was used in ferrets, nonhuman primates, and mice. The vaccination showed complications in the animals regarding their immunity (Tseng et al., 2012). Soon the virus disappeared due to many factors such as summer weather and quarantine of affected individuals. So further studies on vaccine development and clinical trials were stopped (Tseng et al., 2012).No cases were heard or seen after 2004. No more infections were detected. This way the SARS pandemic came to an end (Zumla and Zumla, 2019). SARS-CoV-like viruses find bats to be potential hosts. They have shown a great ability to infect humankind over and again. It signifies that SARS could reemerge (de Wit et al., 2016).

The Middle East respiratory syndrome (MERS-CoV, +ssRNA) was first diagnosed in 2012. A 60-year-old person died because of respiratory illness and multi organ failure in the city of Jeddah, Kingdom of Saudi Arabia. Later, MERS-CoV was identified as the cause of his death. MERS-CoV is a global public health concern due to recurrent nosocomial and community outbreaks. The mortality is very high. During June–July 2015, MERS-CoV has caused outbreak outside of the Arabian Peninsula, infecting Korea. The outbreak indicates the epidemic potential of MERS-CoV. It spreads from person to person (Zumla and Zumla, 2019).There is not enough data from human clinical trials. Although it is essential to find reasonable data to progressively move forward with discovering possible therapeutic measures (de Wit et al., 2016).

Marburg virus (MARV,–ssRNA) was first discovered in 1967 in Europe. Recent reports have been from Eastern Africa and a massive outbreak in Angola in 2005. It is extremely pathogenic and causes hemorrhagic fever. The mortality rates are as high as 90%. MARV outbreak is sporadic and deadly. The poor resources and facilities are incompatible with the diagnosis and treatment of patients. No vaccine or treatment is approved of 100% efficacy. The sudden reemergence of MARV makes testing ethically and logistically challenging for the medical staff (Shifflett and Marzi, 2019).

Ebola virus is a –ssRNA virus that causes the deadly Ebola virus disease. In the past four decades, several recurrences of Ebola has been reported. It began in 2014, started in Africa and spread to several continents of the world, and became a pandemic. The world should pay extra attention to this virus due to several factors like the unique structure of the virus, its infectivity, lethality, and uncontrolled spreading, and most importantly the lack of an effective treatment (Murray et al., 2015). A review of the Ebola virus can be found in Khan et al. (2017).

Lassa fever outbreaks happen due to the Lassa virus which is an ssRNA virus. It is a zoonotic disease of humans. Lassa fever frequently outbreaks in West Africa with almost around 500,000 cases per year and around 10,000 deaths. Infected rodents are the primary cause of infection in humans. The secondary is person-to-person transmission. The disease exists and shows up again and again. No permanent cure has been known yet. Poor medical conditions make it even worse for the people of West Africa (Asogun et al., 2019).

Rift Valley fever virus (RVFV, –ssRNA) is a zoonotic arbovirus. It causes critical health conditions in humans and livestock. Discovered in 1931, very frequent outbreaks have occurred in Africa and the Arabian Peninsula (Clark et al., 2018). Zika virus is a +ssRNA virus. This disease (Jabeen et al., 2019) is caused by a flavivirus. It is mosquito borne. It is a pandemic and public health emergency (Agumadu and Ramphul, 2018). The signs and symptoms are influenza-like illness. Guillain–Barre syndrome occurs in adults and microcephaly in babies born to Zika virus-infected mothers. No effective treatment or therapeutics/medicine/vaccine is available. The major focus of public health is to prevent the spreading of the infection specifically in pregnant women. Despite several studies and knowledge gathered, questions that remain unanswered are about the virus's vectors, its reservoirs, pathogenesis, genetic diversity, and potential synergistic effects of coinfection with other spreading viruses (Agumadu and Ramphul, 2018). Nipah virus (NiV, RNA virus) is a bat-borne pathogen. Almost 20 years ago it was identified in Malaysia first and since then it has continued to spread in regions of South and Southeast Asia. It causes chronic nervous and respiratory illnesses which is life threatening. It is highly infectious. It spreads through infected animals and humans (Aditi and Shariff, 2019). Early and fast diagnoses along with the implementation of control measures are essential to stop the infection. Difficulties in diagnosis and control show up when a new area gets infected with this virus. The mortality rate is high. So far no effective treatment or vaccine is readily available (Aditi and Shariff, 2019). Hendra virus is a –ssRNA belongs to the genus *Henipavirus*. It has great pandemic potential because it is a highly pathogenic zoonotic virus. It was first identified during the 1990s across Australia and Southern Asia having a mortality rate of up to 75%, known to cause severe respiratory and encephalitic disease and remain rare and sporadic. Very less is known about the effect on the nervous system due to Hendra virus infection (Dawes and Freiberg, 2019). Chikungunya virus (+ssRNA) is an alphavirus, mostly transmitted by Aedes mosquitoes into human beings. It first emerged in Tanzania in 1952–1953.In Asia, it was first reported in Bangkok, Thailand, in 1958. The virus continued to spread till 1964. The virus spread appeared in 1970 and declined after six years in 1976. Till 2005 reports of chikungunya have come in (Chhabra et al., 2008). The transmission from an infected mother to her child can occur during the gestation period. The signs and symptoms of Chikungunya include fever, myalgia, headache, and rashes. Since 2004, massive outbreaks in tropical regions have occurred. Severe complications arise as a result of chikungunya infection including neurological disease, which is being recognized increasingly (Mehta et al., 2018). The most recent novel coronavirus disease (COVID-19), caused by severe acute respiratory syndrome coronavirus 2 (SARS-COV-2, +ssRNA virus), is a global public health concern and declared a pandemic by WHO (Qazi et al., 2021). Originated from the Wuhan City, China, it is assumed to be of zoonotic roots.

Human-to-human transmission of COVID-19 infection triggered an outbreak, and isolation and social distancing are prescribed to break the viral spread. Extensive mitigation strategies to control the pandemic have been adopted. The senior citizens, children, and persons with compromised immunity or poor health condition are at higher risk. The search for effective drugs and vaccines against this virus is still ongoing (Rothan and Byrareddy, 2020).

These reemergent viral diseases discussed above pose major combat challenges to the scientific world again and again. The global level of poor health facilities and diagnosis methods makes the patients helpless. The underdeveloped technologies of detection and prognosis are not enough for full-fledged treatments. The most critical part of it is the development of medicine/drugs or vaccines. Safeguarding people from pandemics is very critical. The truth is that even if therapeutics are discovered, it does not assure 100% recovery of all patients. Therefore, the study of viruses has to be dynamic, including various dimensions such as virology, computer science, biology, physiology, biochemistry, physics, metabolism, and toxicology. This need can be fulfilled with the help of viroinformatics about which we will learn now.

6.3 WHAT IS VIROINFORMATICS?

Viroinformatics which sounds akin to bioinformatics is the amalgamation of virology and computer sciences software applications (Figure 6.2). In other words, it is a subarea of bioinformatics that deals with research specific to viruses. It is a newly emergent field after taking into consideration the amount of data and damage bred

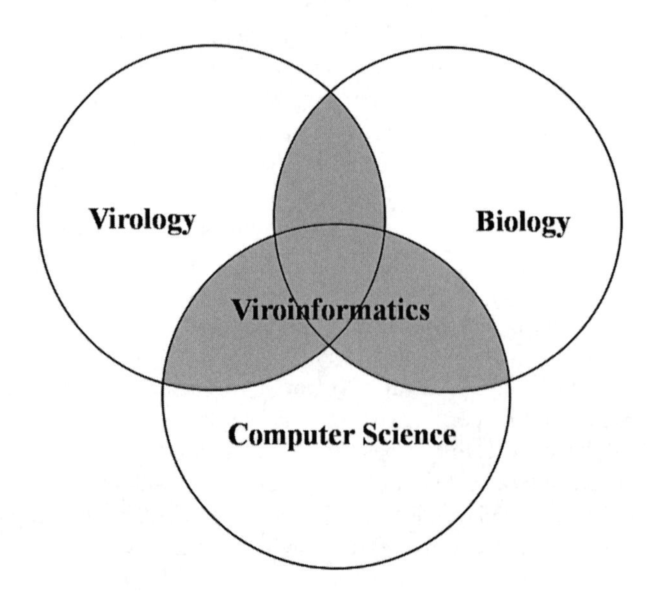

FIGURE 6.2 The interdisciplinary field of viroinformatics.

by it each year (Sharma et al., 2015). Now here, another interesting question arises. Why viroinformatics? Well, the answer is pretty simple. Viroinformatics has been developed for use by the bench virologist and genomic biologists to use it straightforwardly. This you shall discover in the upcoming sections on your own when we learn about the way to use tools.

"Virology" deals with a variety of strains of viruses with distinct biological properties such as genetic organization, replication methods, host range, and interaction mechanism with the host (commensalism, parasitism, ammensalism, and mutualism). The most significant property of the viruses is to mutate its genetic makeup and adapt to host immune responses and therapeutic interventions (Marz et al., 2014; Chang 2015). There are several basic questions in virology that need to be answered including how we can apprehend the evolutionary diversity of viruses and its families in different environments and hosts? How do virus genomes mutate at such a quick pace? Why and how is genetic recombination vital? What is the source of virus origination? How do we recognize the dynamism of gene pool carried by several viruses in different types of ecosystems/ecologies?

Several similar types of queries will give us the insight to know the expansive role of viruses in ecology besides controlling them in infections and pandemics. The great potential of genome sequencing methods and technologies, associated with newly formed tools to handle "big data," facilitates a wide scope to get these doubts answered. Here we should emphasize that questions that remain unanswered in virology need special support and attention of the bioinformatics department. The combined efforts and expertise of bioinformaticians and virologists could bring some revolutionary change (Marz et al., 2014; Chang, 2015).

6.3.1 Viroinformatics-Based Databases and Tools

The European Virus Bioinformatics Centre (EVBC, http://evbc.uni-jena.de/tools/) maintains a list of all the possible virus-based tools and web services available for the virologists and bioinformaticians. Based on the following two categories, databases and tools, the available online viral services can be classified. Besides EVBC list, there are several independent databases and tools available. We will discuss them in a later section.

6.3.1.1 Viroinformatical Databases

Databases are the repository or data storage services. They store multiple similar data in an organized form and are easy to retrieve and read. Table 6.2 provides an insight into the web services/databases related to viruses. The common repositories are eggnog (Huerta-Cepas et al., 2019), EpiFlu (Shu and McCauley, 2017), HCV database group (Kuiken et al., 2005), HSV-1 genome browser (Whisnant et al., 2020), ICTV, IRAM (Almansour et al., 2019), MMRdb (Almansour & Alhagri, 2019), OrthoDB, ViPR (Pickett et al., 2012), HCV portal of ViPR (Zhang et al., 2019), ViralZone (Hulo et al., 2011), VirHostNet (Navratil et al., 2009), VirusesSTRING (Cook et al., 2018), VVR, bNAber, CoVdb, and DengueNet.

TABLE 6.2
The Web Resources/Database of Viroinformatics

Repository/ Database	Description	URL
eggNOG v5.0	An online repository of orthology relationships, functional annotations, and evolutionary history of more than 2,500 viral proteomes belonging to 5,090 organisms. It provides orthologs by sequence similarity, and Fasta sequences can be annotated	http://eggnog5.embl.de/#/app/home Huerta-Cepas et al. (2019)
EpiFlu	Publicly accessible influenza viruses database which contains genetic sequence and clinical and epidemiological data	https://platform.gisaid.org/epi3/frontend#1064eb Shu and McCauley (2017)
HCV	HCV database group presents HCV-related, human-annotated genetic data, via web-accessible search interfaces that provide access to the central database. Provides analysis tools too	http://hcv.lanl.gov Kuiken et al. (2005)
HSV-1	It can be accessed for the complete transcriptome and translatome of herpes simplex virus	https://zenodo.org/record/3465873#.XsRMv6gzZPZ Whisnant et al. (2020)
ICTV	International Committee on Taxonomy of Viruses (ICTV) is for the taxonomy and classification nomenclature of viruses	https://talk.ictvonline.org/taxonomy/
IRAM	Virus capsid data analysis web access. Nucleotide and amino-acid comparisons can be done Via blastp and blastn	https://iram.iau.edu.sa/#/ Almansour et al. (2019)
MMRdb	Database for measles, mumps, and rubella viruses. Allows analysis too	http://mmrdb.org/#/ Almansour and Alhagri (2019)
OrthoDB	For evolutionary annotations and functional ortholog annotations, mapping of data by the user	https://www.orthodb.org/ Pickett et al. (2012)
ViPR	Multiple virus family data are included. Information about the 3D structure of the protein, sequences, host factor, immune epitopes, drugs against viruses, etc. can be looked for. Using SNP, a phylogenetic tree can be constructed. BLAST-based comparative study is done	https://www.viprbrc.org/brc/home.spg?decorator=vipr Zhang et al. (2019)
Hepatitis C virus (HCV)	HCV-based research can be done. Development of HCV therapeutics. Great for analysis. It also has tools for visualization and data mining. Helps in hypothesis generation	https://www.viprbrc.org/brc/home.spg?decorator=flavi_hcv Zhang et al. (2019)

(*Continued*)

TABLE 6.2 (*Continued*)
The Web Resources/Database of Viroinformatics

Repository/ Database	Description	URL
ViralZone	SIB web resource for all viral genus provides molecular and epidemiological information	https://viralzone.expasy.org/ Hulo et al. (2011)
VirHostNet	Knowledgebase which allows proteome-wide analysis of virus–host interaction networks. It also includes information relevant to the COVID-19 pandemic in the March 2020 release	http://virhostnet.prabi.fr/ Navratil et al. (2009)
Viruses STRING	Database for protein–protein interaction between virus–host and virus–virus	http://viruses.stringdb.org/ Cook et al. (2018)
Virus Variation Resource (VVR)	It is a web-based tool for information retrieval, analysis, and visualization of sequences	https://www.ncbi.nlm.nih.gov/ genome/viruses/variation/
bNAber	HIV antibodies neutralizing database	http://bnaber.org
CoVDB	For the coronavirus genome annotated data	http://covdb.microbiology.hku.hk
DengueNet	Dengue fever and dengue hemorrhagic studies and dataset	http://www.who.int/denguenet

6.3.1.2 Viroinformatical Tools

Viroinformatical tools allow the user to perform computational analysis. The tools can be classified into various categories based on the type of applications. Table 6.3 lists the tools and its applications for viral studies are described as follows:

De novo assembly method allows the construction of genome by linking of de novo contigs by combining the scores of nucleotides and aminoacids. Tools namelyAV454 (Assemble Viral 454) (Hennet al., 2012), Genome Detective (Vilsker et al., 2019), SPAdes (Bankevich et al., 2012), V-FAT (Viral Finishing and Annotation Toolkit), VICUNA (Yang et al., 2012), and VrAP (Viral Assembly Pipeline) are de novo assembly based.

Sequencing and annotation tools are Artic, Augustus (Nachtweide and Stanke, 2019), Base-By-Base (Tu et al., 2018), DIGS (Zhu et al., 2018), DinuQ (Lytras et al., 2020), GLUE (Singer et al., 2018), LAMPA (Gulyaeva et al., 2020), PoSeiDon, PriSeT (Hoffmann et al., 2020b), PriSM, Tanoti, VIGOR4 (Wang et al., 2010), VIpower (Sulovari and Li, 2020), ViraMiner (Tampuu et al., 2019), VIRULIGN (Libin et al., 2019), etc. These tools allow performing sequence-based analyses such as sequence similarity, identification, alignment, quantification, amplification, and comparison, while the annotation is done to maintain the descriptions and details associated with the virus genome.

Secondary structure prediction tools such as LRIscan (Fricke and Marz, 2016), RNAalifold (Lorenz et al., 2011), and Silent Mutations (SIM) (Desirò et al., 2019) enable the virologists to predict the secondary structure (alphahelix, betasheets, loops) from the amino-acid sequence of the gene/genome of the virus.

TABLE 6.3
Some of the Popular Web Tools Available for Viral Studies

Application	Tools	Description	URL
De novo assembly	AV454 (Assemble Viral 454)	From a population of viral genomes mixture assembled read data is generated	https://www.broadinstitute.org/viral-genomics/av454 Henn et al. (2012)
	Genome detective	Amino-acids and nucleotide scores are combined, construction of genomes by reference-based linking of *de novo* contigs is done, this is done by an alignment method	https://www.genomedetective.com/ Vilsker et al. (2019)
	SPAdes	The assembly of genomes and mini-metagenomes is done using highly chimeric reads	http://cab.spbu.ru/software/spades/ Bankevich et al.(2012)
	V-FAT (Viral Finishing and Annotation Toolkit)	An automated computational tool used for annotation of *de novo* viral assemblies	https://www.broadinstitute.org/viral-genomics/v-fat
	VICUNA	Assembles ultra-deep sequence data with good fold coverage	https://www.broadinstitute.org/viral-genomics/vicuna Yang et al. (2012)
	VrAP (Viral Assembly Pipeline)	Identifies viruses without sequence homology to known references	https://www.rna.uni-jena.de/research/software/vrap-viral-assembly-pipeline/
Sequencing and annotation	ARTIC	nCoV workflow system	https://github.com/artic-network/fieldbioinformatics
	AUGUSTUS	Multi-genome annotation can be done	http://bioinf.uni-greifswald.de/augustus/ Nachtweide and Stanke(2019)
	Base-By-Base	For viruses with the large genome, comparative analysis	https://4virology.net/virology-ca-tools/base-by-base/ Tu et al. (2018)
	DIGS	Database-integrated genome screening	http://giffordlabcvr.github.io/DIGS-tool/ Zhu et al. (2018) (2018)
	DinuQ	Dinucleotide quantification	https://pypi.org/project/dinuq/ Lytras and Hughes (2020)

(*Continued*)

TABLE 6.3 (*Continued*)
Some of the Popular Web Tools Available for Viral Studies

Application	Tools	Description	URL
	GLUE	For multiple sequence alignment and data, for sequence database creation RDBMS and MySQL can be used	http://glue-tools.cvr.gla.ac.uk/#/home Singer et al. (2018)
	LAMPA	LArge Multidomain Protein Annotator, protein annotation for the large domains of proteins of a virus	https://github.com/Gorbalenya-Lab/hh-suite-notebooks/tree/LAMPA Gulyaeva et al. (2020)
	PriSeT	Discovery of de novo Primer	https://github.com/mariehoffmann/PriSeT Hoffmann et al. (2020b)
	PriSM	Degenerate primers for the amplification and sequencing of short viral genomes	https://www.broadinstitute.org/viral-genomics/prism Yu et al. (2011)
	Tanoti	BLAST-guided reference-based short read aligner	https://www.bioinformatics.cvr.ac.uk/tanoti.php
	VIGOR4	Viral Genome ORF Reader	https://github.com/JCVenterInstitute/VIGOR4 Wang et al. (2010)
	VIpower	Estimating the power of Viral Integration	http://www.uvm.edu/genomics/software/VIpower/live/ Sulovari and Li (2020)
	ViraMiner	Identifying viral genomes in human samples	https://github.com/rega-cev/virulign Tampuu et al. (2019)
	VIRULIGN	Fast codon-correct alignment and annotation of viral genomes	https://github.com/rega-cev/virulign Libin et al. (2019)
	LRIscan	Based on a multiple genome alignment long-range interactions prediction of full viral genomes	https://www.rna.uni-jena.de/supplements/lriscan/ Fricke and Marz (2016)
Secondary structure prediction	RNAalifold	Secondary structures prediction of RNA-aligned sequences	https://www.tbi.univie.ac.at/RNA/RNAalifold.1.html Lorenz et al. (2011)
	SilentMutations (SIM)	Analysis of the effect of multiple point mutations on secondary structures of two interacting viral RNAs	https://github.com/desiro/silentMutations Desirò et al. (2019)

(*Continued*)

TABLE 6.3 (*Continued*)
Some of the Popular Web Tools Available for Viral Studies

Application	Tools	Description	URL
Virus genotyping and diagnosis	ArboTyping	Virus species and genotypesin identification of dengue, zika, and chikungunya	http://krisp.ukzn.ac.za/app/typingtool/virus/ Fonseca et al. (2019)
	ATHLATES	Accurate typing of human leukocyte antigen through exome sequencing	https://www.broadinstitute.org/viral-genomics/athlates Liu et al. (2013)
	CCHFV Primer Checker	Molecular assays efficacy detection and to detect as many CCHFV strains by specific sets of assays to	https://github.com/cesaregruber/CCHFV-PrimerChecker Gruber et al. (2019)
	DisCVR	Rapid viral diagnosis from HTS data	https://bioinformatics.cvr.ac.uk/software/discvr/ Maabar et al. (2019)
	ERVcaller	Detection of polymorphic endogenous retrovirus and other transposable element insertion.	Chen and Li (2019)
	geno2pheno	Using NGS data for detecting the resistance of viral drug. Interpretation is done by a genotypic system	https://ngs.geno2pheno.org/ Döring et al. (2018)
	Purple	A computational framework for the strategic selection of peptides for diagnosing virals using MS-based targeted proteomics	Lechner et al. (2019)
	RotaC2.0	Genotype differentiation of 11 groups A rotavirus segments of genes	Maes et al. (2009)
	TaxIt	A computational working pipeline for untargeted strain-level identification based on MS/MS spectra	https://gitlab.com/rki_bioinformatics Kuhring et al. (2020)
	ViPR	Rotavirus A genotype determination	https://www.viprbrc.org/brc/rvaGenotyper.spg?method=ShowCleanInputPage&decorator=reo#
	V-Pipe	Extraction of genomes of viruses along with the improvement of the diagnostic	https://www.viprbrc.org/brc/rvaGenotyper.spg?method=ShowCleanInputPage&decorator=reo# Carlisle et al. (2019)

(*Continued*)

TABLE 6.3 (*Continued*)
Some of the Popular Web Tools Available for Viral Studies

Application	Tools	Description	URL
Phylogenetic analysis	AdaPatch	Applied in denser and spatially distinct clusters of sites under positive selection on the surface of proteins, and it is targeted on protein structures of viruses of unknown adaptive behavior	https://research.bifo.helmholtz-hzi.de//software/ Tusche et al. (2012)
	AntiPatch	On the protein structure, antigenicity-altering sites are present, used for inference	https://github.com/hzi-bifo/PatchInference Kratsch et al. (2016)
	Antigenic Tree Inference	The phylogenetic tree is inferred by analyzing virus sequences and assigning antigenic distances to reconstructed amino-acid changes on internal branches	https://research.bifo.helmholtz-hzi.de//software/ Steinbrück and McHardy (2012)
	BEAST	The Bayesian phylogenetic and phylodynamic data integration tool	https://beast-dev.github.io/beast-mcmc/ Suchard et al. (2018), Bouckaert et al. (2019)
	EPA-ng	Evolutionary placement algorithm for NGS	https://github.com/Pbdas/epa-ng Barbera et al. (2019)
	Fréchet tree distance measure	Phylo-geographies comparison across multiple trees inferred from the same taxa	https://github.com/hzi-bifo/FrechetTreeDistance Reimering et al. (2018)
	SANTA-SIM	Simulation of viral sequence evolution dynamics under selection and recombination	https://github.com/santa-dev/santa-sim Jariani et al. (2019)
	Sweep Dynamics (SD) plots	Computational identification of selective sweeps	https://github.com/hzi-bifo/SDplots Klingen et al. (2018)
	VICTOR	Virus classification and tree building online resource	https://ggdc.dsmz.de/victor.php Meier-Kolthoff et al. (2017)
	ViCTree	Computes a maximum likelihood phylogeny by automatically selecting new virus sequences from GenBank and generates MSA	https://bioinformatics.cvr.ac.uk/victree/ Modha et al. (2018)

(*Continued*)

TABLE 6.3 (*Continued*)
Some of the Popular Web Tools Available for Viral Studies

Application	Tools	Description	URL
Metagenomics	CAMISIM	Simulating metagenomes and microbial communities	Fritz et al. (2019)
	EPA-ng	Evolutionary placement algorithm for NGS	https://github.com/Pbdas/epa-ng Barbera et al. (2019)
	LiveKraken	Real-time metagenomic classification of Illumina sequencing data	Tausch et al. (2018)
	RIEMS	Metagenomic sequence datasets information extraction	https://www.fli.de/en/institutes/institute-of-diagnostic-virology-ivd/laboratories-working-groups/laboratory-for-ngs-and-microarray-diagnostics/#c4651 Scheuch et al. (2015)
	vConTACT v.2.0	Taxonomic assignment of uncultivated prokaryotic virus genomes	https://bitbucket.org/MAVERICLab/vcontact2/src/master/ Bin et al. (2019)
	VirSorter	Extraction of signal of virus from microbial genomic data	https://github.com/simroux/VirSorter Roux et al. (2015)

Virus genotyping and diagnosis tools allow us to study the detection efficiency of existing molecular assays and to reveal new specific sets of assays to detect. Tools such as ArboTyping (Fonseca et al., 2019), ATHLATES (Liu et al., 2013), CCHFV Primer Checker (Gruber et al., 2019), DisCVR (Maabar et al., 2019), ERVcaller (Chen and Li 2019), geno2pheno (Döring et al., 2018), Purple (Lechner et al., 2019), RotaC2.0 (Maes et al., 2009), TaxIt (Kuhring et al.,2020), ViPR, and V-Pipe (Carlisle et al., 2019) serve the purpose with many other similar functions.

Phylogenetic analysis tools are AdaPatch (Tusche et al., 2012), AntiPatch (Kratsch et al., 2016), Antigenic Tree Inference (Steinbrück and McHardy, 2012), BEAST 1.10 (Suchard et al., 2018) (Bouckaert et al., 2019), EPA-ng (Barbera et al., 2019), Fréchet tree distance measure (Reimering et al., 2018), SANTA-SIM (Jariani et al., 2019), Sweep Dynamics (SD) plots (Klingen et al., 2018), TaxIt, VICTOR (Meier-Kolthoff et al., 2017), and ViCTree (Modha et al., 2018). These tools enable the biologists to find the evolutionary relationships among the different strains of viruses. The taxonomy and lineage can be identified for evolutionary studies and comparative analysis.

Metagenomics tools are critical for simulating metagenomes and microbial communities. Some of the tools are CAMISIM (Fritz et al., 2019), EPA-ng (Barbera et al., 2019), LiveKraken (Tausch et al., 2018), RIEMS (Scheuch et al., 2015), vConTACT v.2.0 (Bin Jang et al., 2019), and VirSorter (Roux et al., 2015).

All of these databases and tools can be used in a systematic scientific manner to obtain insights into a virus. However, the tools are not fully perfect and have a great room for development as per the need of users.

6.4 VIROINFORMATICS: NEED OF THE HOUR

Looking at the present pandemic situations and frequent outbreaks of different viral diseases in different parts of the world, it's high time for a massive collaboration between bioinformatics and viroinformatics. The world researchers have to remarkably think upon this together. However, there have been few bioinformatics research societies that are paying a great attention to viruses (Marz et al., 2014). With certain exceptional cases, viral genomes are mostly poorly annotated. Several computational methods and tools have been proposed and developed that facilitates the analysis and discovery of characteristic features of individual virus families. Due to the small genome size of the viruses, it is possible to sequence a large number of isolates samples. This bulk of sequencing and structural data produced during each pandemic needs specialized methods of analysis, which are partially available today. The currently available technologies for sequencing viral genomes portray great challenges because major steps of the analysis pipeline are not frequently automated and every methodological approach has its specific disadvantages (Marz et al., 2014). However, by amalgamating bioinformatics approaches, tools, and techniques, it could be possible to predict the evolution of the virus in patients based on individual virus population characteristics in the future. Hence, there is an urgent need to develop an integrated working pipeline that combines various processing steps to study viral diversity (Hufsky et al., 2018). Those working pipelines would then help both clinicians and virologists to discover and characterize the underlying disease-causing virus populations (Hufsky et al., 2018).

6.4.1 Future Perspectives of Viroinformatics

Human minds have constantly achieved milestones when it comes to discovery and research. For the first time in history, the genome of a virus was sequenced, that is where probably viroinformatics was born but remained without a name for years. Bioinformaticians are people with understanding of biology and maths. They not only can analyze a high-throughput data or perform a BLAST in a straightforward way but actually develop those system-specific algorithms that work behind these tools.

Clearly, at present just a bunch of virologists are tackling the pandemics and outbreaks. The only accessibility they have is to the laboratories with limited computations. Maybe if they are provided with weapons in the form of powerful data generation viroinformatical tools, they could more efficiently handle the viruses. For that very purpose, EVBC has been set up (Hufsky et al., 2018). It was established with the belief that such a center would produce lots of algorithms and tools and that was what happened in reality too. EVBC has successfully developed viroinformatical tools for detection of viruses from high-throughput sequencing data. It has discovered tools for virus assembly and quasi-species reconstruction. To understand the intra-viral interactions tools have been developed. Virus entry and virus–host interaction analysis tools have been the aim of EVBC too along with the purpose to understand the protein–protein interaction. Efficient phylogenetic analysis tool is in the priority tool list too. But the major concern remains to identify the therapeutic methods. There are no unified databases for viruses. The well-established repository-like NCBI has certain loopholes for virologists such as they are asked to assign the name of a chromosome, which cannot be done for viruses (Hufsky et al., 2018)

The scope for these types of tools development remains wide. The future holds a lot more new discoveries and developments in viroinformatics. As the advancements in bioinformatics would grow, the viroinformatics is also expected to enhance. There is an urgent need to bring virologists and bioinformaticians together. We seriously need young and vibrant scientists with the knowledge of viruses and skills of bioinformatics to bridge the gap between the two of them (Hufsky et al., 2018).

6.5 CONCLUSION

This chapter presents an insight into the basic understanding of the emerging interdisciplinary field of viroinformatics. The thorough list of viral diseases, its causes, and the global effect on health tells us how deadly the viral diseases can be. The ones mentioned in this chapter have created much havoc and panic for years until they were eliminated or controlled to some extent. However, the poor strategies to control the damage persist even in these times of extensive machine learning, artificial intelligence, and "Big data" analysis. The potential of viroinformatics is huge. The use of these tools and databases would enable you to think on that part. The scope of the field is broad. Several aspects should be unfolded soon. The development of software that could predict the alarming situations of pandemic needs to be created, an application that could automatically produce the list of 100% potential drugs just by the structural information of proteins as input would help clinicians and medical teams. Besides these, software or program exclusively should be developed for designing novel drugs for the new emerging viral disease, and that would be a boon. Sincere efforts from all fields' experts have been constantly pushing the boundaries of possibilities. Viroinformatics is still developing and expanding its wings. The way we approach it to solve healthcare issues is a major concern. Viroinformatics can be exclusively used for a proper channel analysis of a specific viral outbreak. Deeper studies of upcoming trends and technologies would guide you to achieve the possibilities.

REFERENCES

Aditi, & Shariff, M. (2019). Nipah virus infection: A review. *Epidemiology and Infection*, 147, e95. https://doi.org/10.1017/S0950268819000086.

Agumadu, V. C., & Ramphul, K. (2018). Zika virus: A review of literature. *Cureus*, 10(7), e3025. https://doi.org/10.7759/cureus.3025.

Almansour, I., & Alhagri, M. (2019). MMRdb: Measles, mumps, and rubella viruses database and analysis resource. *Infection, Genetics and Evolution: Journal of Molecular Epidemiology and Evolutionary Genetics in Infectious Diseases*, 75, 103982. https://doi.org/10.1016/j.meegid.2019.103982.

Almansour, I., Alhagri, M., Alfares, R., Alshehri, M., Bakhashwain, R., & Maarouf, A. (2019). IRAM: virus capsid database and analysis resource. *Database: The Journal of Biological Databases and Curation*, *2019*, baz079. https://doi.org/10.1093/database/baz079.

Asogun, D. A., Günther, S., Akpede, G. O., Ihekweazu, C., & Zumla, A. (2019). Lassa fever: Epidemiology, clinical features, diagnosis, management and prevention. *Infectious Disease Clinics of North America*, *33*(4), 933–951. https://doi.org/10.1016/j.idc.2019.08.002.

Bankevich, A., Nurk, S., Antipov, D., Gurevich, A. A., Dvorkin, M., Kulikov, A. S., Lesin, V. M., Nikolenko, S. I., Pham, S., Prjibelski, A. D., Pyshkin, A. V., Sirotkin, A. V., Vyahhi, N., Tesler, G., Alekseyev, M. A., & Pevzner, P. A. (2012). SPAdes: A new genome assembly algorithm and its applications to single-cell sequencing. *Journal of Computational Biology: A Journal of Computational Molecular Cell Biology*, *19*(5), 455–477. https://doi.org/10.1089/cmb.2012.0021.

Barbera, P., Kozlov, A. M., Czech, L., Morel, B., Darriba, D., Flouri, T., & Stamatakis, A. (2019). EPA-ng: Massively parallel evolutionary placement of genetic sequences. *Systematic Biology*, *68*(2), 365–369. https://doi.org/10.1093/sysbio/syy054.

Bashar, I., Dino, M., Franziska, H., Martin, B., Li, D., Mercier, L. P., Massimo, P., Volker, T., & Manja, M. (2018). A new era of virus bioinformatics. *Virus Research*, *251*. https://doi.org/10.1016/j.virusres.2018.05.009.

Bin Jang, H., Bolduc, B., Zablocki, O., Kuhn, J. H., Roux, S., Adriaenssens, E. M., Brister, J. R., Kropinski, A. M., Krupovic, M., Lavigne, R., Turner, D., & Sullivan, M. B. (2019). Taxonomic assignment of uncultivated prokaryotic virus genomes is enabled by gene-sharing networks. *Nature Biotechnology*, 37(6), 632–639. https://doi.org/10.1038/s41587-019-0100-8.

Bouckaert, R., Vaughan, T. G., Barido-Sottani, J., Duchêne, S., Fourment, M., Gavryushkina, A., Heled, J., Jones, G., Kühnert, D., De Maio, N., Matschiner, M., Mendes, F. K., Müller, N. F., Ogilvie, H. A., du Plessis, L., Popinga, A., Rambaut, A., Rasmussen, D., Siveroni, I., Suchard, M. A., … Drummond, A. J. (2019). BEAST 2.5: An advanced software platform for Bayesian evolutionary analysis. *PLoS Computational Biology*, *15*(4), e1006650. https://doi.org/10.1371/journal.pcbi.1006650.

Carlisle, L. A., Turk, T., Kusejko, K., Metzner, K. J., Leemann, C., Schenkel, C. D., Bachmann, N., Posada, S., Beerenwinkel, N., Böni, J., Yerly, S., Klimkait, T., Perreau, M., Braun, D. L., Rauch, A., Calmy, A., Cavassini, M., Battegay, M., Vernazza, P., Bernasconi, E., … Swiss

HIV Cohort Study (2019). Viral diversity based on next-generation sequencing of HIV-1 provides precise estimates of infection recency and time since infection. *The Journal of infectious diseases*, *220*(2), 254–265. https://doi.org/10.1093/infdis/jiz094.

Chakraborty, I., & Maity, P. (2020). COVID-19 outbreak: Migration, effects on society, global environment and prevention. *The Science of the Total Environment*, *728*, 138882. Advance online publication. https://doi.org/10.1016/j.scitotenv.2020.138882.

Chang, J., 2015. Core services: Reward bioinformaticians. *Nature*, *520*, 151–152. http://dx.doi.org/10.1038/520151a.

Chen, X., & Li, D. (2019). ERVcaller: Identifying polymorphic endogenous retrovirus and other transposable element insertions using whole-genome sequencing data. *Bioinformatics (Oxford, England)*, *35*(20), 3913–3922. https://doi.org/10.1093/bioinformatics/btz205.

Chhabra, M., Mittal, V., Bhattacharya, D., Rana, U., & Lal, S. (2008). Chikungunya fever: A re-emerging viral infection. *Indian Journal of Medical Microbiology*, *26*(1), 5–12. https://doi.org/10.4103/0255-0857.38850.

Clark, M., Warimwe, G. M., Di Nardo, A., Lyons, N. A., & Gubbins, S. (2018). Systematic literature review of Rift Valley fever virus seroprevalence in livestock, wildlife and humans in Africa from 1968 to 2016. *PLoS Neglected Tropical Diseases*, *12*(7), e0006627. https://doi.org/10.1371/journal.pntd.0006627.

Cook, H. V., Doncheva, N. T., Szklarczyk, D., von Mering, C., & Jensen, L. J. (2018). Viruses. STRING: A virus-host protein-protein interaction database. *Viruses*, *10*(10), 519. https://doi.org/10.3390/v10100519.

Dawes, B. E., & Freiberg, A. N. (2019). Henipavirus infection of the central nervous system. *Pathogens and Disease*, *77*(2), ftz023. https://doi.org/10.1093/femspd/ftz023.

de Wit, E., van Doremalen, N., Falzarano, D., & Munster, V. J. (2016). SARS and MERS: recent insights into emerging coronaviruses. *Nature Reviews. Microbiology*, *14*(8), 523–534. https://doi.org/10.1038/nrmicro.2016.81.

Desirò, D., Hölzer, M., IbrahimBashar, B., & Marz, M. (2019). SilentMutations (SIM): A tool for analyzing long-range RNA-RNA interactions in viral genomes and structured RNAs. *Virus Research*, *260*, 135–141. https://doi.org/10.1016/j.virusres.2018.11.005.

Döring, M., Büch, J., Friedrich, G., Pironti, A., Kalaghatgi, P., Knops, E., Heger, E., Obermeier, M., Däumer, M., Thielen, A., Kaiser, R., Lengauer, T., & Pfeifer, N. (2018). geno2pheno[ngs-freq]: A genotypic interpretation system for identifying viral drug resistance using next-generation sequencing data. *Nucleic Acids Research*, *46*(W1), W271–W277. https://doi.org/10.1093/nar/gky349.

Durmuş, S., & Ülgen, K. Ö. (2017). Comparative interactomics for virus-human protein-protein interactions: DNA viruses versus RNA viruses. *FEBS Open Bio*, *7*(1), 96–107. https://doi.org/10.1002/2211-5463.12167.

Enquist, L. W., & Editors of the Journal of Virology (2009). Virology in the 21st century. *Journal of Virology*, *83*(11), 5296–5308. https://doi.org/10.1128/JVI.00151-09.

Fonseca, V., Libin, P., Theys, K., Faria, N. R., Nunes, M., Restovic, M. I., Freire, M., Giovanetti, M., Cuypers, L., Nowé, A., Abecasis, A., Deforche, K., Santiago, G. A., Siqueira, I. C., San, E. J., Machado, K., Azevedo, V., Filippis, A., Cunha, R., Pybus, O. G., … de Oliveira, T. (2019). A computational method for the identification of Dengue, Zika and Chikungunya virus species and genotypes. *PLoS Neglected Tropical Diseases*, *13*(5), e0007231. https://doi.org/10.1371/journal.pntd.0007231.

Fricke, M., &Marz, M. (2016). Prediction of conserved long-range RNA-RNA interactions in full viral genomes. *Bioinformatics (Oxford, England)*, *32*(19), 2928–2935. https://doi.org/10.1093/bioinformatics/btw323.

Fritz, A., Hofmann, P., Majda, S., Dahms, E., Dröge, J., Fiedler, J., Lesker, T. R., Belmann, P., DeMaere, M. Z., Darling, A. E., Sczyrba, A., Bremges, A., & McHardy, A. C. (2019). CAMISIM: simulating metagenomes and microbial communities. *Microbiome*, *7*(1), 17. https://doi.org/10.1186/s40168-019-0633-6.

Gelderblom, H/R. Structure and Classification of Viruses. In: Baron, S., editor. *Medical Microbiology*. 4th edition. Galveston (TX): University of Texas Medical Branch at Galveston; 1996. https://www.ncbi.nlm.nih.gov/books/NBK8174/.

Gruber, C., Bartolini, B., Castilletti, C., Mirazimi, A., Hewson, R., Christova, I., Avšič, T., Grunow, R., Papa, A., Sánchez-Seco, M. P., Kopmans, M., Ippolito, G., Capobianchi, M. R., Reusken, C., & Di Caro, A. (2019). Geographical variability affects CCHFV detection by RT-PCR: A tool for in-silico evaluation of molecular assays. *Viruses*, *11*(10), 953. https://doi.org/10.3390/v11100953.

Gulyaeva, A. A., Sigorskih, A. I., Ocheredko, E. S., Samborskiy, D. V., & Gorbalenya, A. E. (2020). LAMPA, LArge multidomain protein annotator, and its application to RNA virus polyproteins. *Bioinformatics (Oxford, England)*, *36*(9), 2731–2739. https://doi.org/10.1093/bioinformatics/btaa065.

Henn, M. R., Boutwell, C. L., Charlebois, P., Lennon, N. J., Power, K. A., Macalalad, A. R., Berlin, A. M., Malboeuf, C. M., Ryan, E. M., Gnerre, S., Zody, M. C., Erlich, R. L., Green, L. M., Berical, A., Wang, Y., Casali, M., Streeck, H., Bloom, A. K., Dudek, T., Tully, D., … Allen, T. M. (2012). Whole genome deep sequencing of HIV-1 reveals the impact of early minor variants upon immune recognition during acute infection. *PLoS Pathogens*, *8*(3), e1002529. https://doi.org/10.1371/journal.ppat.1002529.

Hoffmann, M., Kleine-Weber, H., Schroeder, S., Krüger, N., Herrler, T., Erichsen, S., Schiergens, T. S., Herrler, G., Wu, N. H., Nitsche, A., Müller, M. A., Drosten, C., & Pöhlmann, S. (2020a). SARS-CoV-2 cell entry depends on ACE2 and TMPRSS2 and is blocked by a clinically proven protease inhibitor. *Cell*, *181*(2), 271–280.e8. https://doi.org/10.1016/j.cell.2020.02.052.

Hoffmann, M., Monaghan, M.T., & Reinert, K. (2020b).PriSeT: Efficient de novo primer discovery. *bioRxiv*2020.04.06.027961. https://doi.org/10.1101/2020.04.06.027961.

Huerta-Cepas, J., Szklarczyk, D., Heller, D., Hernández-Plaza, A., Forslund, S. K., Cook, H., Mende, D. R., Letunic, I., Rattei, T., Jensen, L. J., von Mering, C., & Bork, P. (2019). eggNOG 5.0: A hierarchical, functionally and phylogenetically annotated orthology resource based on 5090 organisms and 2502 viruses. *Nucleic Acids Research*, *47*(D1), D309–D314. https://doi.org/10.1093/nar/gky1085.

Hufsky, F., Ibrahim, B., Beer, M., Deng, L., Mercier, P. L., McMahon, D. P., Palmarini, M., Thiel, V., & Marz, M. (2018). Virologists-Heroes need weapons. *PLoS Pathogens*, *14*(2), e1006771. https://doi.org/10.1371/journal.ppat.1006771.

Hulo, C., de Castro, E., Masson, P., Bougueleret, L., Bairoch, A., Xenarios, I., & Le Mercier, P. (2011). ViralZone: A knowledge resource to understand virus diversity. *Nucleic Acids Research*, *39*(Database issue), D576–D582. https://doi.org/10.1093/nar/gkq901.

Jabeen, A., Ahmad, N. & Raza, K. (2019). *Differential Expression Analysis of ZIKV Infected Human RNA Sequence Reveals Potential Genetic Biomarkers*. Lecture Notes in Bioinformatics, Springer, 11465: 1–12. https://doi.org/10.1007/978-3-030-17938-0_26.

Jariani, A., Warth, C., Deforche, K., Libin, P., Drummond, A. J., Rambaut, A., Matsen Iv, F. A., & Theys, K. (2019). SANTA-SIM: simulating viral sequence evolution dynamics under selection and recombination. *Virus Evolution*, *5*(1), vez003. https://doi.org/10.1093/ve/vez003.

Khan, F.N., Qazi, S., Tanveer, K. & Raza, K.. (2017). *A Review on the Antagonist Ebola: A Prophylactic Approach. Biomedicine & Pharmacotherapy*, Elsevier, 96: 1513–1526. https://doi.org/10.1016/j.biopha.2017.11.103.

Klingen, T. R., Reimering, S., Loers, J., Mooren, K., Klawonn, F., Krey, T., Gabriel, G., & McHardy, A. C. (2018). Sweep dynamics (SD) plots: Computational identification of selective sweeps to monitor the adaptation of influenza A viruses. *Scientific Reports*, *8*(1), 373. https://doi.org/10.1038/s41598-017-18791-z.

KoprowskiH. (1993). Yesterday, today and tomorrow in virology. *Intervirology*, *35*(1–4), 9–15. https://doi.org/10.1159/000150291.

Kratsch, C., Klingen, T. R., Mümken, L., Steinbrück, L., & McHardy, A. C. (2016). Determination of antigenicity-altering patches on the major surface protein of human influenza A/H3N2 viruses. *Virus Evolution*, *2*(1), vev025. https://doi.org/10.1093/ve/vev025.

Kuhring, M., Doellinger, J., Nitsche, A., Muth, T., & Renard, B. Y. (2020). TaxIt: An iterative computational pipeline for untargeted strain-level identification using MS/MS spectra from pathogenic single-organism samples. *Journal of Proteome Research*, 10.1021/acs.jproteome.9b00714. Advance online publication. https://doi.org/10.1021/acs.jproteome.9b00714.

Kuiken, C., Yusim, K., Boykin, L., & Richardson, R. (2005). The Los Alamos hepatitis C sequence database. *Bioinformatics (Oxford, England)*, *21*(3), 379–384. https://doi.org/10.1093/bioinformatics/bth485.

Lechner, J., Hartkopf, F., Hiort, P., Nitsche, A., Grossegesse, M., Doellinger, J., Renard, B. Y., & Muth, T. (2019). Purple: A computational workflow for strategic selection of peptides for viral diagnostics using MS-based targeted proteomics. *Viruses*, *11*(6), 536. https://doi.org/10.3390/v11060536.

Libin, P., Deforche, K., Abecasis, A. B., & Theys, K. (2019). VIRULIGN: Fast codon-correct alignment and annotation of viral genomes. *Bioinformatics (Oxford, England)*, *35*(10), 1763–1765. https://doi.org/10.1093/bioinformatics/bty851.

Liu, C., Yang, X., Duffy, B., Mohanakumar, T., Mitra, R. D., Zody, M. C., & Pfeifer, J. D. (2013). ATHLATES: Accurate typing of human leukocyte antigen through exome sequencing. *Nucleic Acids Research*, *41*(14), e142. https://doi.org/10.1093/nar/gkt481.

Lorenz, R., Bernhart, S. H., HönerZuSiederdissen, C., Tafer, H., Flamm, C., Stadler, P. F., & Hofacker, I. L. (2011). ViennaRNA Package 2.0. *Algorithms for Molecular Biology: AMB*, *6*, 26.https://doi.org/10.1186/1748-7188-6-26.

Lytras,S., & Hughes, J. (2020).Synonymous dinucleotide usage: A codon-aware metric for quantifying dinucleotide representation in viruses. *Viruses*, *12*, 4, 462.

Maabar, M., Davison, A. J., Vučak, M., Thorburn, F., Murcia, P. R., Gunson, R., Palmarini, M., & Hughes, J. (2019). DisCVR: Rapid viral diagnosis from high-throughput sequencing data. *Virus Evolution*, *5*(2), vez033. https://doi.org/10.1093/ve/vez033.

Maes, P., Matthijnssens, J., Rahman, M., & Van Ranst, M. (2009). RotaC: A web-based tool for the complete genome classification of group A rotaviruses. *BMC Microbiology*, *9*, 238. https://doi.org/10.1186/1471-2180-9-238.

Marz, M., Beerenwinkel, N., Drosten, C., Fricke, M., Frishman, D., Hofacker, I. L., Hoffmann, D., Middendorf, M., Rattei, T., Stadler, P. F., & Töpfer, A. (2014). Challenges in RNA virus bioinformatics. *Bioinformatics (Oxford, England)*, *30*(13), 1793–1799. https://doi.org/10.1093/bioinformatics/btu105.

Mehta, R., Gerardin, P., de Brito, C., Soares, C. N., Ferreira, M., & Solomon, T. (2018). The neurological complications of chikungunya virus: A systematic review. *Reviews in Medical Virology*, *28*(3), e1978. https://doi.org/10.1002/rmv.1978.

Meier-Kolthoff, J. P., & Göker, M. (2017). VICTOR: genome-based phylogeny and classification of prokaryotic viruses. *Bioinformatics (Oxford, England)*, *33*(21), 3396–3404. https://doi.org/10.1093/bioinformatics/btx440

Modha, S., Thanki, A. S., Cotmore, S. F., Davison, A. J., & Hughes, J. (2018). ViCTree: An automated framework for taxonomic classification from protein sequences. *Bioinformatics (Oxford, England)*, *34*(13), 2195–2200. https://doi.org/10.1093/bioinformatics/bty099.

MurrayM. J. (2015). Ebola virus disease: A review of its past and present. *Anesthesia and Analgesia*, *121*(3), 798–809. https://doi.org/10.1213/ANE.0000000000000866.

Nachtweide, S., & Stanke, M. (2019). Multi-genome annotation with AUGUSTUS. *Methods in Molecular Biology (Clifton, N.J.)*, *1962*, 139–160. https://doi.org/10.1007/978-1-4939-9173-0_8.

Navratil, V., de Chassey, B., Meyniel, L., Delmotte, S., Gautier, C., André, P., Lotteau, V., & Rabourdin-Combe, C. (2009). VirHostNet: A knowledge base for the management and the analysis of proteome-wide virus-host interaction networks. *Nucleic Acids Research, 37*(Database issue), D661–D668. https://doi.org/10.1093/nar/gkn794.

Pickett, B. E., Sadat, E. L., Zhang, Y., Noronha, J. M., Squires, R. B., Hunt, V., Liu, M., Kumar, S., Zaremba, S., Gu, Z., Zhou, L., Larson, C. N., Dietrich, J., Klem, E. B., & Scheuermann, R. H. (2012). ViPR: An open bioinformatics database and analysis resource for virology research. *Nucleic Acids Research*, *40*(Database issue), D593–D598. https://doi.org/10.1093/nar/gkr859.

Qazi, S., Sheikh, K., Faheem, M., Khan, A., & Raza, K. (2021). A coadunation of biological and mathematical perspectives on the pandemic COVID-19: A review. Coronaviruses, vol. 2. https://doi.org/10.2174/2666796702666210114110013

Raza, K. (2020). Computational Intelligence Method in COVID-19: Surveillance, Prevention, Prediction, and Diagnosis. *Studies in Computational Intelligence (SCI)*, Springer, 923.

Raza, K., Maryam, Qazi, S. (2020). *An Introduction to Computational Intelligence for COVID-19: Surveillance, Prevention, Prediction, and Diagnosis*. Studies in Computational Intelligence (SCI), Springer, 923.

Reimering, S., Muñoz, S., & McHardy, A. C. (2018). A Fréchet tree distance measure to compare phylogeographic spread paths across trees. *Scientific Reports*, *8*(1), 17000. https://doi.org/10.1038/s41598-018-35421-4.

Rothan, H. A., & Byrareddy, S. N. (2020). The epidemiology and pathogenesis of coronavirus disease (COVID-19) outbreak. *Journal of Autoimmunity*, *109*, 102433. https://doi.org/10.1016/j.jaut.2020.102433.

Roux, S., Enault, F., Hurwitz, B. L., & Sullivan, M. B. (2015). VirSorter: Mining viral signal from microbial genomic data. *Peer Journal*, *3*, e985. https://doi.org/10.7717/peerj.985.

Scheuch, M., Höper, D., & Beer, M. (2015). RIEMS: A software pipeline for sensitive and comprehensive taxonomic classification of reads from metagenomics datasets. *BMC Bioinformatics*, *16*(1), 69. https://doi.org/10.1186/s12859-015-0503-6.

Sharma, D., Priyadarshini, P., & Vrati, S. (2015). Unraveling the web of viroinformatics: computational tools and databases in virus research. *Journal of Virology*, *89*(3), 1489–1501. https://doi.org/10.1128/JVI.02027-14.

Shifflett, K., & Marzi, A. (2019). Marburg virus pathogenesis - differences and similarities in humans and animal models. *Virology Journal*, *16*(1), 165. https://doi.org/10.1186/s12985-019-1272-z.

Shu, Y., & McCauley, J. (2017). GISAID: Global initiative on sharing all influenza data - from vision to reality. *Euro Surveillance: Bulletin Europeen sur les Maladies Transmissibles = European Communicable Disease Bulletin*, *22*(13), 30494. https://doi.org/10.2807/1560-7917.ES.2017.22.13.30494.

Singer, J. B.,Thomson, E. C., McLauchlan, J., Hughes, J., & Gifford, R. J. (2018). GLUE: A flexible software system for virus sequence data. *BMC Bioinformatics*, *19*(1), 532. https://doi.org/10.1186/s12859-018-2459-9

Steinbrück, L., & McHardy, A. C. (2012). Inference of genotype-phenotype relationships in the antigenic evolution of human influenza A (H3N2) viruses. *PLoS Computational Biology*, *8*(4), e1002492. https://doi.org/10.1371/journal.pcbi.1002492.

Suchard, M. A., Lemey, P., Baele, G., Ayres, D. L., Drummond, A. J., & Rambaut, A. (2018). Bayesian phylogenetic and phylodynamic data integration using BEAST 1.10. *Virus Evolution*, *4*(1), vey016. https://doi.org/10.1093/ve/vey016.

Sulovari, A., & Li, D. (2020). VIpower: Simulation-based tool for estimating power of viral integration detection via high-throughput sequencing. *Genomics*, *112*(1), 207–211. https://doi.org/10.1016/j.ygeno.2019.01.015.

Tampuu, A., Bzhalava, Z., Dillner, J., & Vicente, R. (2019). ViraMiner: Deep learning on raw DNA sequences for identifying viral genomes in human samples. *PloS One*, *14*(9), e0222271. https://doi.org/10.1371/journal.pone.0222271.

Tausch, S. H., Strauch, B., Andrusch, A., Loka, T. P., Lindner, M. S., Nitsche, A., & Renard, B. Y. (2018). LiveKraken--real-time metagenomic classification of illumina data. *Bioinformatics (Oxford, England)*, *34*(21), 3750–3752. https://doi.org/10.1093/bioinformatics/bty433.

Tseng, C. T., Sbrana, E., Iwata-Yoshikawa, N., Newman, P. C., Garron, T., Atmar, R. L., Peters, C. J., & Couch, R. B. (2012). Immunization with SARS coronavirus vaccines leads to pulmonary immunopathology on challenge with the SARS virus. *PloS One*, *7*(4), e35421. https://doi.org/10.1371/journal.pone.0035421.

Tu, S. L., Staheli, J. P., McClay, C., McLeod, K., Rose, T. M., & Upton, C. (2018). Base-by-base version 3: New comparative tools for large virus genomes. *Viruses*, *10*(11), 637. https://doi.org/10.3390/v10110637.

Tusche, C., Steinbrück, L., & McHardy, A. C. (2012). Detecting patches of protein sites of influenza A viruses under positive selection. *Molecular Biology and Evolution*, *29*(8), 2063–2071. https://doi.org/10.1093/molbev/mss095.

Vilsker, M., Moosa, Y., Nooij, S., Fonseca, V., Ghysens, Y., Dumon, K., Pauwels, R., Alcantara, L. C., VandenEynden, E., Vandamme, A. M., Deforche, K., & de Oliveira, T. (2019). Genome Detective: An automated system for virus identification from high-throughput sequencing data. *Bioinformatics (Oxford, England)*, *35*(5), 871–873. https://doi.org/10.1093/bioinformatics/bty695.

Wang, S., Sundaram, J. P., & Spiro, D. (2010). VIGOR, an annotation program for small viral genomes. *BMC Bioinformatics*, *11*, 451. https://doi.org/10.1186/1471-2105-11-451.

Wang, W., Simmonds, J., Pan, Q., Davidson, D., He, F., Battal, A., Akhunova, A., Trick, H. N., Uauy, C., & Akhunov, E. (2018). Gene editing and mutagenesis reveal inter-cultivar differences and additivity in the contribution of TaGW2 homoeologues to grain size and weight in wheat. *Theoretical and applied genetics*. *131*(11), 2463–2475. https://doi.org/10.1007/s00122-018-3166-7

Whisnant, A. W., Jürges, C. S., Hennig, T., Wyler, E., Prusty, B., Rutkowski, A. J., L'hernault, A., Djakovic, L., Göbel, M., Döring, K., Menegatti, J., Antrobus, R., Matheson, N. J., Künzig, F., Mastrobuoni, G., Bielow, C., Kempa, S., Liang, C., Dandekar, T., Zimmer, R., … Dölken, L. (2020). Integrative functional genomics decodes herpes simplex virus 1. *Nature Communications*, *11*(1), 2038. https://doi.org/10.1038/s41467-020-15992-5.

Yang, X., Charlebois, P., Gnerre, S., Coole, M. G., Lennon, N. J., Levin, J. Z., Qu, J., Ryan, E. M., Zody, M. C., & Henn, M. R. (2012). De novo assembly of highly diverse viral populations. *BMC Genomics*, *13*, 475. https://doi.org/10.1186/1471-2164-13-475.

Yu, Q., Ryan, E. M., Allen, T. M., Birren, B. W., Henn, M. R., & Lennon, N. J. (2011). PriSM: A primer selection and matching tool for amplification and sequencing of viral genomes. *Bioinformatics (Oxford, England)*, *27*(2), 266–267. https://doi.org/10.1093/bioinformatics/btq624.

Zhang, Y., Zmasek, C., Sun, G., Larsen, C. N., & Scheuermann, R. H. (2019). Hepatitis C virus database and bioinformatics analysis tools in the virus pathogen resource (ViPR). *Methods in Molecular Biology (Clifton, N.J.)*, *1911*, 47–69. https://doi.org/10.1007/978-1-4939-8976-8_3.

Zhu, H., Dennis, T., Hughes, J., & Gifford, R. J. (2018). Database-integrated genome screening (DIGS): Exploring genomes heuristically using sequence similarity search tools and a relational database. *bioRxiv*.

Zumla, A., & Hui, D. (2019). Emerging and reemerging infectious diseases: Global overview. *Infectious Disease Clinics of North America*, *33*(4), xiii–xix. https://doi.org/10.1016/j.idc.2019.09.001.

7 Toxin Databases and Healthcare Applications

Sushmita Baishnab and Subrata Sinha
Dibrugarh University

Arabinda Ghosh
Gauhati University

Ashwani Sharma
International Computational Center

Surabhi Johari
Institute of Management Studies University Courses

CONTENTS

7.1 INTRODUCTION

Toxin is a toxic substance made up of small molecules, peptides, or proteins that develop inside living cells or organisms and can cause disease. Ludwig Brieger named such components as toxins, synonymously derived from the word poisonous (Brieger, 1887). Biotoxins are toxins generated spontaneously by microorganisms, including lower microorganisms, fungi (mycotoxins),bacteria, dinoflagellates, and algae and higher plants and animals as well *viz.* phytotoxins and zootoxins that differ widely in their toxicity range (Cavaillon, 2018).

Biotoxins, unlike biological agents, are toxic organism materials and are inanimate and of no value to living beings, but they are believed for their mysterious jobs within the metabolism of the animals. It is known that many phytotoxins protect the host against certain species, particularly insects. Animals secrete defense molecules that are released either extracellular or are concentrated in some tissues, such as spines or fangs. Spiders and a few snakes bear poison as an aid in battling the prey and often use an analogous defensive poison. There are four types of toxic entities generally, chemical, biological, physical, and radiation, and their toxicity can be defined as the degree to which an organism can be affected by a chemical substance or by a specific combination of drugs (Dolan et al., 2010).

Several species generate the complex peptide that can be bunches of neurotoxin compounds present in the venoms and has significant natural purposes. These are highly selective neurotoxins that are used as a good range of ion channels and receptors and may represent effective lead compounds for novel medicines such as analgesics, anticancer drugs, and neurological disorder medicines (Pitschmann and Hon, 2016).

The purpose of this chapter is to create a central repository for all the toxic compounds and create a database. This will mention information about the origin of toxins from various resources, use of those toxins in various applications such as the discovery of new drugs or vaccines, how they act as a substitute for replacing the existing drug, active pharmaceutical ingredients (API) for the synthesis of new active drug molecules, or direct implication in the treatment of the disease. Therefore, this chapter will provide insight for the biologist getting acquaintance and information for different categories of toxins and using them for their therapeutic experiments.

7.1.1 Plant Toxin

Plant toxins are examples of small molecules that are structurally complex and arise from secondary metabolism in plants. Such toxins are of plant secondary metabolites that are deleterious or lethal to humans and animals, such as phenolic substances, terpenes, alkaloids, steroids, and other compounds. Since time immemorial, the castor plant (*Ricinus communis L.)* has been used in the traditional medicine of the ancient Mediterranean and Eastern cultures (Chandra et al., 2012).In addition, it is still used worldwide in folk medicine. Ricinus toxicity is manifested in varieties of molecules such as (a) ricin protein, which belongs to the family of the ribosome-inactivating protein, having been used for many bio-crimes; (b) castor bean is used as a laxative and in curing abscess disease treatment for baldness (Polito et al., 2019).

Additionally, oilseed's antiphlogistic activity is also found to be associated with acute macrophage toxicity that helps in generating inflammatory cytokines. Also, ricin's antipathogenic activity may facilitate lesion healing, thus acting as one of the potent molecules in the treatment of skin ailment. Castor seed toxicity started to be studied in the late nineteenth century when its oilseed and seed pulp were used in drug preparations for oral ingestion recommended in small quantities (Peigneur and Tytga, 2018; Polito et al., 2019).

In traditional Chinese medicine, castor seeds were used for anthelmintic activity and in the treatment for ulcers and chronic wounds. In Ayurveda, the castor plant is used to treat rheumatic disorders, constipation, gastropathy, diarrhea, nausea, ascites, bronchitis, cough, skin diseases, colic, and lumbago (Patel et al., 2013).

7.1.2 Bacterial Toxin

In 1888, Emile Roux and Alexandre Yersin at Pasteur Institute discovered the first bacterial toxin called diphtheria toxin, produced from *Corynebacterium diphtheria* from sterile filtrates (Alouf, 2006). Soon after the diphtheria toxin, two other major toxins were also discovered: tetanus toxin (1890) and botulinum toxin (1896) produced by *Clostridium tetani* and *Clostridium botulinum*, respectively (Escargueil et al., 2018). Behring and Kitasato (1890) made mice and rabbits immune to the toxic effects of tetanus and diphtheria toxins in Berlin by inoculating the animals with small doses of toxin preparations attenuated by "Gram liquor" (iodine). The sera of the "inoculated" creatures explicitly killed the poisons while at the same time moving into the assortments of local creatures being tested with the local poisons. This experiment showed the assembly of neutralizing antibodies for the primary time and contributed to the serotherapy event and the advent of experimental immunology. In 1884, Emile Roux suggested the enormous variety of horse-insusceptible sera (immunizing agents) being developed for diphtheria serotherapy (Popoff, 2018).

7.1.3 Animal Toxin

Animal toxins are proteins, peptides, and small molecular compounds that can be broadly classified into two major categories—Group 1: venomous species that possess a complex venom system comprising a mixture of proteins and peptide toxins, and Group 2: species that acquire and accumulate small molecular and toxic metabolites as toxins from their environment while retaining relative tolerance to toxic effects (Mebs,2002).These toxins are observed in poison-dart frogs, New Guinean Pitohui birds, and African crested rats. In the mid of the last century, intensive study of biochemical poisons began, though some were discovered much earlier within the venoms of snakes. According to the PubMed database, there are around 1,000 articles published annually on venoms, and thus the total number of publications is enriched up to 40,000 (Sitprija and Suteparak, 2008) that have a focus on snake venoms. Selective selection and long-term coevolution have produced intense development, high specificities, and immense molecular diversity for animal toxins. Venom toxins are regarded as invaluable and effective pharmacological testing devices and contemporary clinical therapeutics (Johannes, 2019).

7.2 TOXINS IN HEALTHCARE

7.2.1 Drug Discovery

The use of toxins in modern translational medicines began in 1940 with the introduction of tubocurarine as a muscle relaxant and in anesthesia. Researchers are following novel techniques like high-throughput screening for developing new drugs from animal venom as the sources of potential therapies for clinical use (Harvey, 2014). On the other hand, plant toxins that are structurally diverse small molecules, such as alkaloids terpenes, steroids, and phenolic compounds derived from the plant secondary metabolism, are major lead molecules for modern drug discovery. In addition to secondary metabolites, toxic lectins, a plant protein, are utilized for disease diagnosis and as candidates to develop anticancer drugs. The synergy of "omics" and systems biology with ethnomedicine and studies of plant endophytes are exciting approaches for the discovery of new drugs from natural sources (Günthardt et al., 2018).

7.2.2 Cosmetics

Cosmetology is an emerging field of science that involves technology and utilizes compounds for cosmetics that are of biological origin extracted from the reservoir of bacteria, fungi, and algae (Gupta and Sharma, 2017). Botulinum toxin is a neurotoxic protein produced by the bacterium *Clostridium Botulinum* that can be utilized to treat painful muscle spasms. Neurotoxin A, another toxin discovered in 1950, can be used as a muscle relaxant in cosmetic therapy by inhibiting the release of acetylcholine at the neuromuscular junction (Popoff, 2018).

7.3 TOXINS DATABASES

7.3.1 Toxic Exposome Database T3DB

(a) *Database Description:* There are several online toxin repositories, one such example being Toxin and Toxin Target Database (T3DB) (Lim et al., 2010; Wishart et al., 2015). The T3DB, or the Toxic Exposome Database soon to be referred to, may be a unique bioinformatics resource that combines detailed toxin data with comprehensive toxin target information. The database is overcrowded with 3,678 toxins and their trivial names. Additionally, it contains chemicals, pesticides, narcotics, and food contaminants that are related to 2,073 corresponding target contaminants. There are 42,374 toxins, the target associations for toxins. The database is available athttp://www.t3db.ca/.

(b) *Input and Search Criteria of T3DB:*

  (i) *Structure Search:* In this case, the input is a chemical structure that can be drawn in the given interface, and based on any of the three options viz., similarity, substructure, and exact, we can obtain toxin structures from the database. We can also search all toxins structures by molecular weight (by specifying the range of molecular weight).

(ii) *Text Query:* In this type of search we can search the database by specifying text terms using Boolean logic (AND/OR/NOT) operator.
(iii) *Sequence Search:* In this type of search, we can enter DNA/amino-acid sequence in FASTA format; we can also specify BLAST parameters (cost to open a gap, penalty for mismatch, expectation value, cost to extend a gap, and reward for the match).
(iv) *Toxin Advanced Search:* It is a versatile search system in which we can search the database based on any or all of the 35 fields of the database, some of the fields are Toxin name, Cellular location, Tissue location, and Biological role.
(v) *Spectra Search:* This search consists of the following searches—LC–MS Search, LC–MS/MS Search, GC–MS Search,1D NMR Search, and 2D NMR Search:
   a. *LC–MS Search:* In this search, we Input Query Masses (Da) and search toxins by adjusting the following three attributes, that is, Ion Mode, Adduct Type, Molecular Weight Tolerance ±
   b. *LC–MS/MS Search:* In this search, we input one mass(m/z) and intensity corresponding to one peak per line and we search toxins by adjustment of the following attributes—Parent Ion Mass (Da),Parent Ion Mass Tolerance, Mass/Charge (m/z) Tolerance, Ionization Mode, CID Energy.
   c. *GC–MS Search:* In this search, we input one mass (m/z) and intensity corresponding to one peak per line by adjustment of the attribute Peak Tolerance.
   d. *1D NMR Search:* In this search, we input one chemical shift per line by adjustment of the attribute Spectra Library and Tolerance.
   e. *2D NMR Search:* In this search, we input one cross-peak chemical shift per line with numbers separated by space by adjustment of the following three attributes—Spectra Library, X-axis Tolerance ± (ppm),and Y-axis Tolerance ± (ppm).

7.3.2 Database Kalium

(a) *Database Description:* Kalium is a comprehensive collection of the potassium channel polypeptide ligands. This is an open-access tool and all of the information provided here is compiled manually. The current release presents information about potassium channel ligands from other organisms such as snake, scorpion, and spider (Kuzmenkov et al., 2016). The database is available at https:/kaliumdb.org/.
(b) *Input and Search Criteria of Kalium:* We can obtain toxin data from Kalium by selecting any of six buttons representing organism group(shown in top left corner of figure 2);it filters data in the main table according to the organism groups (viz., snakes, scorpions, spiders, sea anemones, cone snails, and miscellaneous) selected by the user. We can perform BLAST or CLUSTAL alignment by clicking the checkboxes on the left side of the general table and we can export data in csv/text format for selected polypeptides in a text file too.

7.3.3 Animal Toxin Databases (ATDB)

(a) *Database Description:* The ATDB is a uni-database and website (http://protchem.hunnu.edu.cn/toxin) containing more than 3,235 UniProtKB/Swiss-Prot and TrEMBL animal toxins and associated toxin databases as well as published literature. ATDB is intended to store chemical structures and annotation data of all animal toxins and introduces a new system of conserved, standardized terminology to standardize functional annotations of toxins. It may be the most extensive database of toxins, which now includes 3,235 peptide toxins with 5 small molecules of 379 species. It provides a user-friendly web interface for querying, visualizing, and analyzing the data. Users can query ATDB with ATDB's toxin ID, protein name, gene name, macromolecule ID, and keyword (He et al., 2008). The database is available at http://protchem.hunnu.edu.cn/toxin.

(b) *Input and Search Criteria of ATDB:* ATDB proposes three different systems to find toxins: "Search," "Browse," and "Ontology" tabs. Numerous options are offered to refine searches. Using the web interface, users can query with toxin ID, protein or gene name, nucleic acid ID, and keywords. There is also "Advanced text search" for a complex query, which allows users to issue queries based on species group, taxonomic ID, domain name, the number of disulfide bridges, target and sequence length, and molecular weight. We can narrow down or expand our results by using AND or operators. We can also find toxin homologs by BLAST.

7.3.4 Toxin Plants Phytotoxin Database (TPTT)

(a) *Database Description:* The TPPT database may be a multipurpose, freely available database that provides information on 844 poisonous plants in Switzerland and Central Europe and 1,586 of their phytotoxins. They provide biological information on plants, their spatial distribution, and their toxicity to humans and animals with chemical information including identification information, molecular structure, as well as physicochemical properties expected insilico, and human and aquatic toxicity (Barbara et al., 2018). The database is available athttps://www.agroscope.admin.ch/agroscope/en/home/publications/apps/tppt.html.

(b) *Input and Search Criteria of TPPT:* The TPPT database is an offline database available in two different formats: SQLite version and MS-Excel version. We can use the TPPT database to develop a nice web interface, and hence it is the discretion of the developer what input or search criteria they want to include in their web interface.

7.3.5 ArachnoServer Spider Toxin Database

(a) *Database Description:* ArachnoServer is a manually curated database containing information on the sequence, three-dimensional structure, and biological activity of protein toxins derived from spider venom. The database is available at http://www.arachnoserver.org/advancedsearch.html.

(b) *Input and Search Criteria of ArachnoServer Spider Toxin Database:* Archano server database web interface consists of tabs like Search, Browse, BLAST, SpiderP, and Tox Note. In Search Tab, we can search Toxin details by 32 attributes under Toxin Metadata, Biological Activity, Protein information, Source species taxonomy, Literature information, and Mature toxin mass. In the browse tab, the toxins can be browsed by ontology like Araneave Taxonomy, Molecular Targets, Posttranslation Modification, and Phyletic Specificity.

In the BLAST tab, we can input sequences in FASTA format. The query sequence type can be selected as Protein or DNA, and we can BLAST against nucleic sequences or peptide sequences using either blastx or blastp program. We can filter query sequence either DUSTS with blastn or SEG with others. The SpiderP tab consists of a user-friendly interface, in which we can input protein sequence in FASTA format. It predicts the location (or absence) of propeptides in spider toxins.

ToxlNote tab consists of multiple predictions and evaluation tools. It is the spider toxin annotation and evaluation facility, it is a bioinformatics pipeline that aims to fast track the analysis of venom-gland transcriptomes generated by next-generation sequencing projects.

7.3.6 Toxin Antitoxin Database (TADB)

(a) *Database Description:* TADB2.0 is an updated database that offers detailed information on the toxin-antitoxin (TA) loci of bacterial type II. The database is updated relative to the previous iteration and thus the current data schema is used. It identified 6,193 type II TA loci in 870 replicas of bacteria and archaea, including 105 experimentally validated TA loci, with the aid of text mining and manual curing (Xie et al., 2018). The database is available at http:/bioinfo-mml.sjtu.edu.cn/TADB2/.

(b) *Input and Search Criteria of TADB 2.0 Database:* In TADB 2.0 we can browse TA locus pair, toxin, and antitoxin by the organism. We can search for toxins based on Species, TA family, Gene/Protein. There are three tools(TA finder-prediction of Type II TA loci in the bacterial genome sequence,WU-BLAST2 search—against the toxin/antitoxin genes/proteins in TADB, and Linkage to RASTA-Bacteria-a web-based tool for identifying TA loci in prokaryotes)available in TADB 2.0

7.3.7 RISCTOX Database

(a) *Database Description:* RISCTOX is a hazardous substance created to provide simple, structured, and succinct information on the threats to health and the environment posed by chemicals in goods commonly used or treated by businesses. This database comprised over 100,000 chemical agents in files with sufficient details. Nonetheless, the database also contains usable health and environmental risk information. The mere fact that a material is not included in RISCTOX does not mean that it is harmless, but rather that we cannot obtain information about its hazards or that there is no such knowledge. The database is available at https://risctox.istas.net/en/.

(b) *Input and Search Criteria of RISCTOX Database:* We can search toxic and hazardous substances by name of the substance or CAS/EC/Index No either by exact name or by a part of the name. We can obtain substance identification information like Chemical name, identification numbers like CAS number, EC EINECS number, and also additional information like molecular formula. In addition to that, we can also obtain information like environmental effects.

7.3.8 DBETH Bacterial Exotoxins for Humans

(a) *Database Description:* This database is restricted to exotoxins generated by bacteria that are mostly pathogenic to humans. The toxins included are classified into 24 distinct classes. DBETH's goal is to provide a comprehensive database for pathogenic bacteria exotoxins in humans. DBETH also offers its users a platform to identify potential exotoxin by homology-based or analogous methods from sequence data. Using the BLAST tool against toxin sequences or running HMMER against toxin domains detected by DBETH from human pathogenic bacterial exotoxins, users can identify potential exotoxin by pairwise alignment in a homology-based approach. DBETH uses a machine learning approach to identify potential exotoxins in analogy-based approach (Chakraborty et al., 2011).The database is available at http://www.hpppi.iicb.res.in/btox/.

(b) *Input and Search Criteria of DBETH Database:* We can browse the DBETH database by Pathogen name, browse Toxin by Toxin Mechanism, and by Toxin Activity. We can search (a) Toxin Domains by Input Sequence in FASTA format, (b) Toxin Sequences by Input Sequence to run BLASTP against Human Pathogenic Bacterial Toxins, (c) Toxin Structures by Input Sequence to run BLASTP against Human Pathogenic Bacterial Toxin Structure Sequences (it accepts FASTA format or PDB Atom Coordinate file), and (d) Toxin Prediction-Machine Learning Approach to Sequence Classification using Support Vector Machine. It accepts sequence in FASTA format. DBETH also allows plain text search.

7.3.9 Super Toxic Database

(a) *Database Description:* SuperToxic is a database, primarily designed for pharmacists, biochemists, and medical scientists, but also for researchers working in cognate disciplines. It provides access to information about toxic compounds (names, synonyms, and structures). SuperToxic predicts the toxicity of compounds, informs about potential targets in biochemical pathways, and shows potential binding partners. The database also includes links to suppliers, where the compound can be obtained for further investigation (Schmidt et al., 2009). The database is available at http://bioinf-services.charite.de/supertoxic/.

(b) *Input and Search Criteria of Super Toxic Database:* In the SuperToxic database, there are four options available, that is, Toxin search, Structure search, Property search, and Browse database. The toxin can be searched by anyone of the following attributes, that is, Name, Pathway, CAS Number, NSC Number, and Formula. In Structure search, we can input structure either by uploading a mol file or by drawing our structure with Marvin-Sketch plugged into the interface. In Property search, we can specify the lower bound and upper bound for the following attributes—mol weight, atoms, rings, bonds, rotatable bonds, log p, H-bond donors, H-bond acceptors, and toxicity value. In the browse Toxin facility, we can browse all toxins alphabetically or we can download all toxins in csv format.

7.3.10 TAS Mania Bacterial Toxin-Antitoxin (TA Database)

(a) *Database Description:* Bacterial TA systems are attributed to biological functions such as phage defense, plasmid maintenance, and virulence persistence. The TAS mania offers an in-depth annotation of TA loci during a very broad bacterial genome database which is a tool of critical value to the microbiology community. The TAS mania includes the creation of experimental approaches, the identification of new TA groups, and the analysis of TA loci information. This is a vital tool for the microbiology community. The TAS mania involves the design of experimental strategies, the discovery of new TA families, and comparative analysis of the content of TA loci. Also, this database facilitates phylogeny and/or phenotypes analysis such as pathogenicity, persistence, stress resistance, associated host types). In this database, TA annotations map the core bacterial genome but also its plasmids, where applicable (Akarsu et al., 2019).The database is available at https://shiny.bioinformatics.unibe.ch/apps/tasmania/.

(b) *Input and Search Criteria of the TAS mania Database:* We can search the TAS mania database by selecting the first letter of the species from the dropdown list given in the web interface, then it prompts us to type or select the desired species name. It gives us the option to select one or more columns like mm_E_value, TAS_info, operon_structure, molecule_desc, tasmania_hmm_cluster_id, pfam_desc, and protein_sequence which gets displayed. We can select e-value to filter the putative hits operons. We can either download all data or e-value filtered data.

7.4 CONCLUSION

The database for toxins is a unique bioinformatics resource that combines detailed data on toxins with comprehensive target information. These databases focus on providing toxicity mechanisms and target proteins for every toxin. The dual information, in which toxin and toxin target records are interactively linked in both directions, makes them unique for therapeutic applications. This chapter provides deep information about the uses of toxins and their application which is a basic important ingredient for the API development of some drugs. Also, these toxins could be used as a vaccine against microbial and viral infections. This chapter helps the biologist to have collective resources of the toxins and their application so that they can design their experiments using these different categories of toxins and use them for their therapeutic experiments.

ACKNOWLEDGMENTS

The authors thank the Director, Centre for Biotechnology and Bioinformatics, Dibrugarh University, and Director, Institute of Management Studies University Courses Ghaziabad, for providing insightful comments and suggestions in designing the manuscript.

REFERENCES

Akarsu H, Bordes P, Mansour M, Bigot DJ, Genevaux P, Falquet L. (2019). TASmania: A bacterial toxin-antitoxin systems database. *PLoS Comput. Biol.* 15(4):e1006946.

Alouf JE (2006). A 116-year story of bacterial protein toxins (1888–2004): from "diphtheritic poison" to molecular toxinology. In *The Comprehensive Sourcebook of Bacterial Protein Toxins* (Third Edition). J. Alouf and M. Popoff (Eds.). Elsevier, Amsterdam, Netherlands. pp 3–21.

Barbara FG, Hollender J, Hungerbühler K, Scheringer M, Bucheli TD. (2018). Comprehensive toxic plants–phytotoxins database and its application in assessing aquatic micropollution potential. *J. Agric. Food. Chem.* 66(29):7577–7588.

Behring E, Kitasato S. (1890). Ueber das zustandekommen der diphtherie-immunität und der tetanus-immunität bei thieren. *Deutschen Medicinischen Wochenschrift* 49: 1113–1114

Brieger L. (1987). Knowledge of the aetiology of tetanus and comments on cholera red. *Dtsch. Med. Wochenschr.* 13:303–305.

Cavaillon J. (2018). Historical links between toxinology and immunology. *Pathog Dis.* 76(3). doi: 10.1093/femspd/fty019.

Chakraborty A, Ghosh S, Chowdhary G, Maulik U, Chakrabarti S. (2011). DBETH: A database of bacterial exotoxins for human. *Nucleic Acids Res.* 40:615–620.

Chandra SJ, SandhyaS, Vinod R, David B, Komera S, Chaitanya RSNAKK. (2012). Plant toxins-useful and harmful effects. *Hygeia J.D. Med.* 4(1):79–90.

Dolan LC, Matulka RA, Burdock GA. (2010). Naturally occurring food toxins. *Toxins (Basel).*2(9): 2289–2332.

Escargueil CR, Lemichez E, Popoff MR. (2018).Variability of botulinum toxins: Challenges and opportunities for the future. *Toxins (Basel).*10(9): 374.

Günthardt BF, Hollender J, Hungerbühler K, Scheringer M, Bucheli TD. (2018). Comprehensive toxic plants–phytotoxins database and its application in assessing aquatic micropollution potential. *J. Agric. Food Chem.* 66: 7577–7588.

Gupta VK, Sharma B. (2017). Forensic applications of Indian traditional toxic plants and their constituents. *Forensic Res. Criminol. Int. J.* 4(1):27–32.

Harvey AL. (2014). Toxins and drug discovery. *Toxicon*92:193–200.

He QY, He QZ, Deng XC, Yao L, Meng E, Liu ZH, Liang SP. (2008). ATDB: A uni-database platform for animal toxins. *Nucleic Acids Res.* 36:293–297.

Johannes AE. (2019). Structurally robust and functionally highly versatile—c-type lectin (-related) proteins in snake venoms. *Toxins (Basel).*11(3):1–25.

Kuzmenkov AI, Krylov NA, Chugunov AO, Grishin EV, Vassilevski AA. (2016). Kalium: A database of potassium channel toxins from scorpion venom. *Database.*2016: baw056 doi:10.1093/database/baw056.

Lim E, Pon A, Djoumbou Y, Knox C, Shrivastava S, Guo AC, Neveu V, Wishart DS. (2010). T3DB: a comprehensively annotated database of common toxins and their targets. *Nucleic Acids Res.* 38:781–786.

Mebs D. (2002). *Venomous and Poisonous Animals: A Handbook for Biologists, Toxicologists and Toxinologists, Physicians and Pharmacists.* Boca Raton, FL: CRC Press, pp.360.

Patel S, Nag MK, Daharwal SJ, Singh MR, Singh D. (2013). Plant toxins: an overview. *Res. J. Pharmacol. Pharmacodyn.* 5(5):283–288.

Peigneur S, Tytgat J. (2018). Toxins in drug discovery and pharmacology. *Toxins (Basel).*10(3):2–4.

Pitschmann V, Hon Z. (2016). Military importance of natural toxins and their analogs. *Molecules.* 21(5):3–20.

Polito L, Bortolotti M, Battelli MG, Calafato G, Bolognesi A. (2019). Ricin: An ancient story for a timeless plant toxin. *Toxins(Basel).*11(6):2–16.

Popoff MR. (2018). "Bacterial toxins" section in the journal toxins: a fantastic multidisciplinary interplay between bacterial pathogenicity mechanisms, physiological processes, genomic evolution, and subsequent development of identification methods, efficient treatment, and prevention of toxigenic bacteria. *Toxins (Basel).*10(1):1–3.

Schmidt U, Struck S, Gruening B, et al. (2009) SuperToxic: A comprehensive database of toxic compounds. *Nucleic Acids Res.* 37(Database Issue):D295–D299.

Sitprija V, Suteparak S. (2008). Animal toxins: An overview. *Asian Biomed.* 2:451–457.

Wishart D, Arndt D, Pon A, Sajed T, Guo AC, Djoumbou Y, Knox C, Wilson M, Liang Y, Grant J, Liu Y, Goldansaz SA, Rappaport SM. (2015). T3DB: The toxic exposome database. *Nucleic Acids Res.* 43:928–934.

Xie Y, Wei Y, Shen Y, Li X, Zhou H, Tai C, Deng Z, Ou HY. (2018). TADB 2.0: An updated database of bacterial type II toxin–antitoxin loci. *Nucleic Acids Res.* 46:749–753.

Part III

Medical Image Processing and Other Healthcare Applications

8 Lossless Medical Image Compression Using Hybrid Block-Based Algorithm for Telemedicine Application

Lenin Fred A. and L. R. Jonisha Miriam
Mar Ephraem College of Engineering and Technology

S. N. Kumar
Amal Jyothi College of Engineering

Ajay Kumar H.
Mar Ephraem College of Engineering and Technology

Parasuraman Padmanabhan and Balázs Gulyás
Cognitive Neuroimaging Centre (CONIC), Lee Kong Chian School of Medicine, Nanyang Technological University

CONTENTS

8.1 INTRODUCTION

The importance of image compression techniques escalated in our advanced communication technology. Reduction in the number of bits is needed for storage and transmission of an image without any loss of information. Diagnosis of diseases with the help of DICOM (digital imaging and communications in medicine) images and their storage plays a vital role in the medical field but it obsesses large bandwidth. For medical applications, DICOM images have to be transferred to diverse destinations. Principal factors of medical images should be preserved during compression achieving a higher compression ratio (CR) and the ability to decode compressed images with original quality are the major issues in medical image compression techniques [1]. Moreover, the reconstructed image yields limited redundancy with good human visual perception at the receiver end. Therefore, it leads to efficient image storage transmission in telemedicine, teleradiology, and real-time teleconsultation.

Advancements in image compression have been suggested in response to expanding requirements in the field of medical imaging. JPEG 2000 is one of the best image compression algorithm [2–4] which utilizes a new coding method called embedded block coding with optimized truncation of embedded bitstreams. Information loss in the compression algorithm is entirely avoided [5]. It [6,7] achieves efficient coding in both lossy and lossless operation. The compression technique in [8–11] reduces utilization power and computational complexity by combining data extension and lifting scheme methods. Integer wavelet transform (IWT) has been represented [12,13] mainly for targeting attention against the joint problem of optimal factorizations and precise representation effects.

In [14,15], IWT implementation leads to adequate lossless and lossy compression performance. A novel compression algorithm was proposed in [16] that achieves a high CR and low latency. The medical image was splitted into nonoverlapped blocks according to its nature and applies first-order polynomial representation to discard the redundancy between neighboring pixels, thereby the compression was performed with a better value of compression rate [17].

The techniques used in [18] consume low power when combining data extension procedure into the lifting-based Discrete Wavelet Transform (DWT) core in an embedded extension algorithm. Analysis of [19] indicates high compression performance achieved with outstanding reconstruction. Compression techniques [20] extract quantitative information and provide original image visuals to humans with a reduced amount of data. Selective image compression technique [21] describes that the regions of interest are compressed in a lossless manner, whereas image regions containing irrelevant information are compressed in a lossy manner.

The classical Vector Quantization (VQ) [22] with a stationary wavelet transform was used for the volumetric image compression. For entropy coding, Huffman and arithmetic coding are used. The stationary wavelet transform generates efficient results, when compared with discrete wavelet transform, lifting transform, and discrete cosine transform [23,24]. The efficiency of the VQ compression technique was improved by the incorporation of a fuzzy S tree [25]. Compression quality was enhanced when the curvelet transform was coupled with the VQ algorithm [26,27].

Wavelet transform was coupled with a prediction model and robust results were produced for CT/MR and US images [28]. A hybrid compression model comprising VQ with Artificial Bee Colony (ABC) and Genetic Algorithm (GA) was used for the optimum codebook selection [29]. The lossless prediction model with wavelet transform was proposed in [30,31], efficient results are produced, when compared with the JPEG standard.

The author compared the performance of spatial tree partitioning (STW) and Set Partitioning in Hierarchical Trees (SPIHT) in terms of mean square error (MSE), peak signal-to-noise ratio (PSNR), CR, and size for various levels ranging from 1 to 8. Result shows that STW outperforms better than SPIHT [13,32,33]. The amalgamation of context and hyper prior models is compared with JPEG, JPEG2000, and BPG for determining the performance concerning the rate distortion parameter on the Kodak dataset. It concludes that this combined model provides good rate distortion [34].

Both KITTI Stereo and KITTI General datasets have been utilized for deep image compression which focuses mainly on decoder side information and yields state-of-the-art performance by the way of Pearson correlation score for various bits per pixel [35]. Deep convolutional autoencoder (CAE) was demonstrated on the Kodak image set to achieve excellent PSNR. Moreover, CAE is followed by Principal Component Analysis (PCA) for enhancing the coding efficiency [36]. Deep learning algorithm is a prominent machine learning algorithm and plays a vital role in the classification [37,38] and compression of medical images [39]. The simulated annealing algorithm was used for the codebook selection in the Contextual Vector Quantization; it's a lossy algorithm but yields fruitful results in terms of reconstructed image quality, when compared with VQ, CVQ, and classical compression algorithms [40]. The least square-based prediction algorithm is a lossless technique and it yields efficient results when compared with JPEG lossless, CVQ, and BAT-VQ algorithms [41].

This chapter is organized as follows. Section 8.2 explains the technical background such as wavelet transform and Hadamard transform. Section 8.3 describes the materials and methods with detailed work. Section 8.4 shows the comparison of IWT-lossless Hadamard transform (LHT)-Huffman, IWT-LHT-Arithmetic, and JPEG lossless encoding techniques. Finally, Section 8.5 gives the conclusion.

8.2 TECHNICAL BACKGROUND

8.2.1 Overview of Wavelet Transform

The development of the wavelet domain [42,43] is progressing well and plays a vital role in image compression. Adequately, wavelets have been used as an extraordinary tool in the areas of signal processing, physics, astronomy, and image processing.

Generally, traditional wavelet transforms have a floating-point algorithm. Hence, it cannot be reconstructed perfectly and it leads to rounding errors. Due to computer-finite word length and higher storage space, complexity in computation is high. To overcome these drawbacks, a finite filter is used in every IWT to lift steps with a finite number.

In many cases, representation of filter output coefficients in wavelet transform is the floating point. In consideration with the representation of input images as matrices and integer values, although the integers of filtered output consist no longer, the result will be in quantization loss. The mapping of the input integer image is essential in lossless coding, with an integer wavelet representation.

The implementation of wavelet transform can be with a filter bank which is considered as a sub-band transform. Figure 8.1 describes the general block scheme of the one-dimensional biorthogonal wavelet transform [44,45]. In this work, 5/3 LWT is used for prediction and is updated in equations as follows.

$$V(2y+1)^{\text{prediction}} = \begin{cases} U(2y+1) - \dfrac{U(2y)+U(2y+2)}{2} & \text{Normal} \\ U(2y+1) - U(2y) & \text{Odd end} \end{cases} \tag{8.1}$$

$$V(2y)^{update} = \begin{cases} U(2y) + \dfrac{V(2y+1)+1}{2} & \text{Normal} \\ U(2y) + V(2y_1) + V(2y+1) + 2 & \text{Even end} \end{cases} \tag{8.2}$$

where V is the 1-D transformation result of U, the prediction is represented as normal in equation and updation of symmetric periodic extension, the boundary extension problem is resolved with odd end and even end in the embedded extension. Transformation occurs while renewing the first data of the image, the even end as

$$V(0) = U(0) + \frac{V(1)+1}{2} \tag{8.3}$$

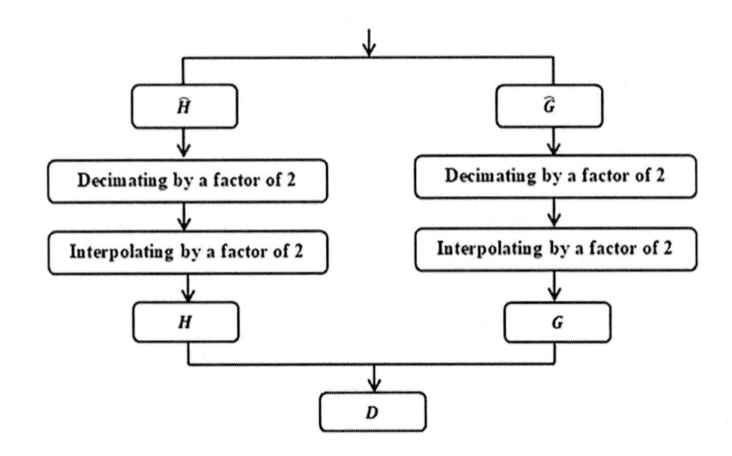

FIGURE 8.1 Basic filter bank for the biorthogonal wavelet transform.

$V(1)$ is estimated from the normal in the above equation.

$$V(1) = U(1) - \frac{U(0) + U(2)}{2} \tag{8.4}$$

According to equation $U(0)$, $U(1)$ and $U(2)$ performed, and the row transformation and column transformation need three data to operate. The transformation can be finished and 2-D IWT can be performed only with three rows of image data. Here, there is no need for entire image data, as it buffers only two rows of data.

8.2.2 Overview of Hadamard Transform

Hadamard and Haar's transformation has a significant computational advantage over the previous Discrete Fourier Transform (DFT), Discrete Cosine Transform (DCT), and Discrete Sine Transform (DST). Addition and subtraction are the main operations involved in the transformation of unitary matrices. It does not use any of the multiplication operations. The multiplication process is time consuming but we need to save time.

The order of Hadamard unitary matrix n is the $N \times N$ matrix $= 2^2$, which brings about the following iteration order.

$$H_n = H_1 \otimes H_{n-1} \tag{8.5}$$

where

$$H_1 = \frac{1}{\sqrt{2}} \begin{bmatrix} 1 & 1 \\ 1 & -1 \end{bmatrix} \tag{8.6}$$

Hadamard transform is very simple and fast and the correlation in the block is eliminated with 2-D LHT [46–48]. The accumulation of the DC component is more precise and for further removal of correlation, this will be very useful. The 2-D LHT is denoted as,

$$L = HSH^T \tag{8.7}$$

where

$$H = H^T = \begin{bmatrix} 1 & 1 & 1 & 1 \\ 1 & 1 & -1 & -1 \\ 1 & -1 & -1 & 1 \\ 1 & -1 & 1 & -1 \end{bmatrix} \tag{8.8}$$

$$\text{Det}(H) = 16 \tag{8.9}$$

$$S' = H^T LH = H^T HST^T H = 16s \tag{8.10}$$

The above expression indicates that the output coefficients are 16 times larger than the input coefficients.

8.3 MATERIALS AND METHODS

The medical images are taken from the Siemens Prisma machine. Datasets are obtained from Cognitive Neuroimaging Centre, Lee Kong Chian School of Medicine, Nanyang Technological University, Singapore, and the study of MRI images of human subjects for research work is approved.

Three important transformations, namely 2-D IWT, 2-D LHT, and DC Prediction, are used to discard the redundancy in both IWT-LHT-Huffman and IWT-LHT-Arithmetic encoder compression algorithms. Finally, to achieve lossless compression, IWT-LHT-Huffman encoder or IWT-LHT-arithmetic encoder is used for further truncation of the code. The block diagram and flow chart are shown in Figures 8.2 and 8.3, respectively.

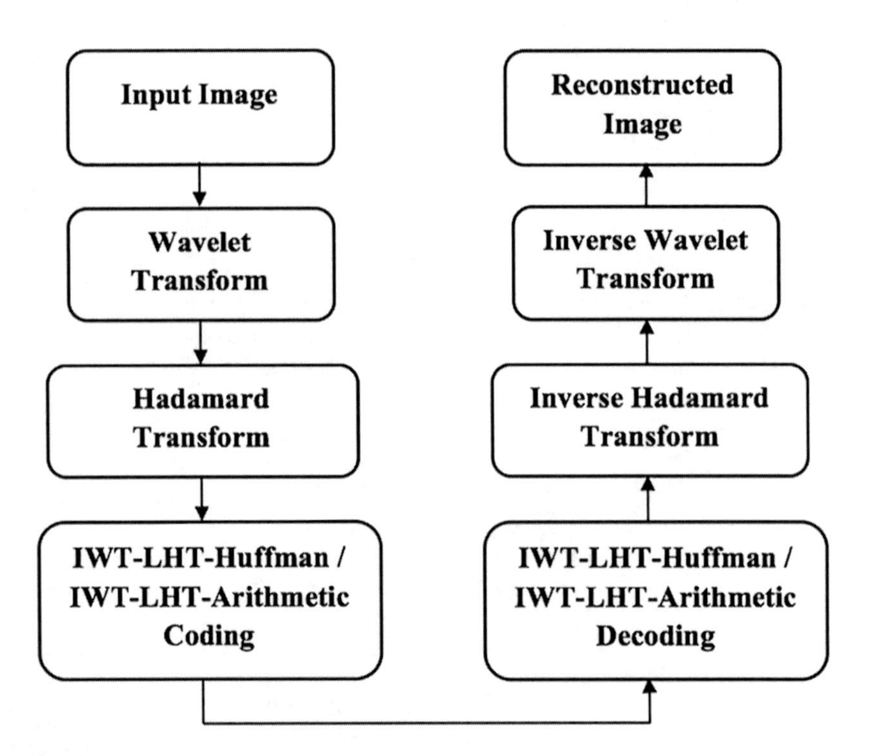

FIGURE 8.2 Block diagram of the integer wavelet transform (IWT)-lossless Hadamard transform (LHT)-Huffman/Arithmetic compression coder.

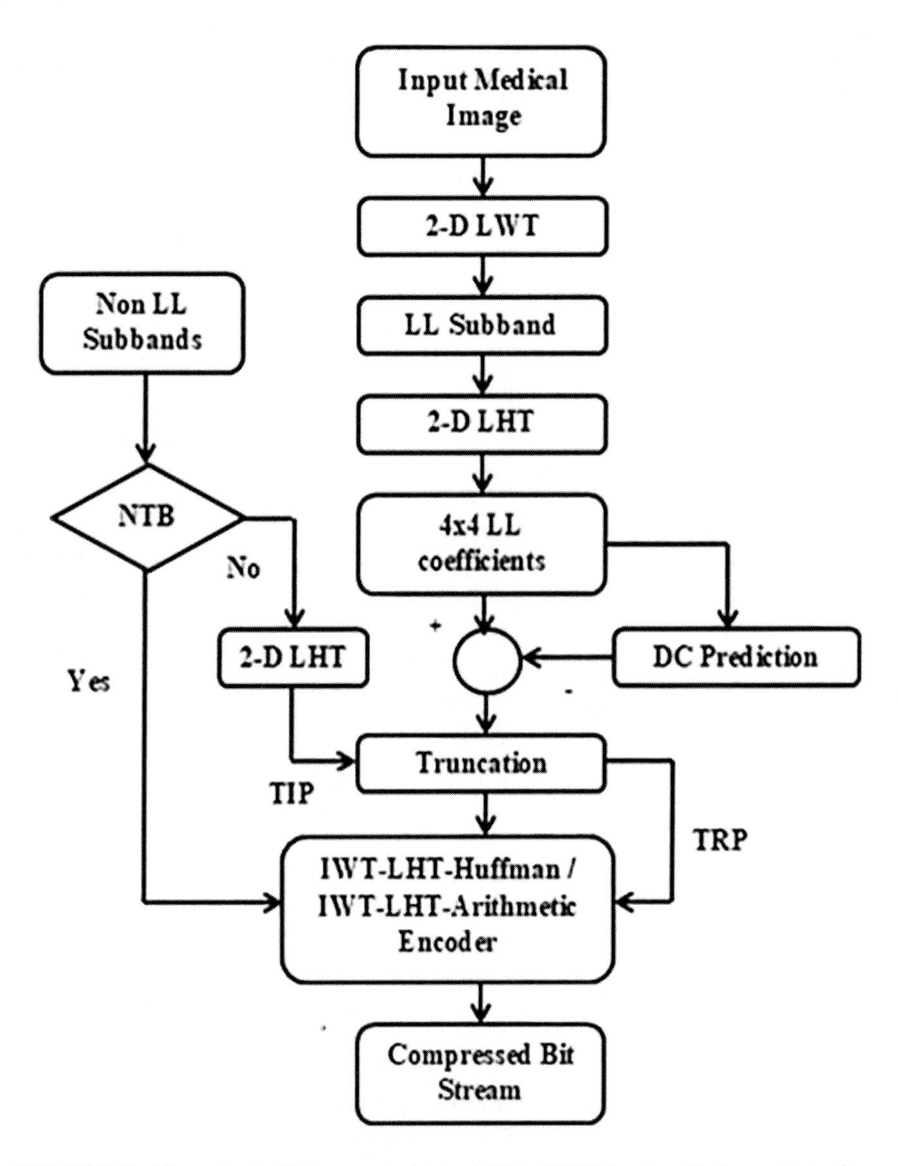

FIGURE 8.3 Flow chart of integer wavelet transform (IWT) lossless Hadamard transform (LHT)-Huffman/Arithmetic compression coder.

Initially, the input MRI image is decomposed by using 2-D LIWT to form a set of band-limited components called sub-bands. Only 4×4 macroblock of LL sub-bands can be processed by 2-D LHT and all the coefficients are transformed into binary form.

In the 2-D LHT coefficient, the error is reduced by clearing the first 3 bits and the last 4 bits (truncated residue part (TRP)). The uncleared 8 bits are termed as the truncated integer part (TIP). From the TRP coefficients, two least significant bit coefficients are denoted as TRPARR (truncated residue part after redundancy reduction). The resultant matrix of the image is given as.

$$\begin{bmatrix} 93 & 112 & 105 & 101 \\ 114 & 244 & 193 & 239 \\ 112 & 227 & 184 & 170 \\ 111 & 210 & 120 & 159 \end{bmatrix} \tag{8.11}$$

$$-2D\ LHT \rightarrow \begin{bmatrix} 2494 & -48 & -296 & -430 \\ -92 & -102 & 82 & 48 \\ -472 & 130 & 130 & 124 \\ -286 & 16 & -8 & 198 \end{bmatrix}$$

$$\begin{bmatrix} 155 & -6 & -30 & -18 \\ -3 & -7 & 8 & 1 \\ -19 & 5 & 8 & -1 \\ -27 & 3 & 7 & 12 \end{bmatrix} - \begin{bmatrix} -2 & 4 & -8 & 2 \\ 0 & -6 & 2 & 0 \\ -8 & 2 & 2 & -8 \\ 2 & 0 & -4 & 6 \end{bmatrix}$$

$$\xrightarrow{\text{Redundant removed}} \begin{bmatrix} 155 & -6 & -30 & -18 \\ -3 & -7 & 8 & 1 \\ -19 & 5 & 8 & -1 \\ -27 & 3 & 7 & 12 \end{bmatrix} + \begin{bmatrix} 3 & 1 & 2 & 0 \\ 0 & 2 & 0 & 0 \\ 2 & 0 & 0 & 2 \\ 0 & 0 & 3 & 1 \end{bmatrix} \tag{8.12}$$

More zeros are favored after truncation so that the adjustment can be followed as,

$$I_{i,j} = I_{i,j} + 1 \; I_{i,j} < 0 \tag{8.13}$$

8.3.1 DC Prediction for LL Sub-Band Coding

In DC prediction (DCP), the quantization and coding are applied to each of the frequency component sub-bands. This quantization and coding are more precisely applied to each of the sub-bands. For the LL sub-band, TIP and TRPARR are the two parts which have to be encoded. The implementation of the DCP technique is simple and is interpreted in Figure 8.4.

8.3.2 Transformation for Non-LL Sub-Band Coding

In the non-LL sub-band validation section, the coefficients of 2-D IWT are authenticated for NTB based on the threshold value before applying it to the IWT-LHT-Huffman encoder. Based on the entropy value, coefficients are again transformed by 2-D LHT to discard unwanted data and the transformed coefficients are encoded after completion of truncation.

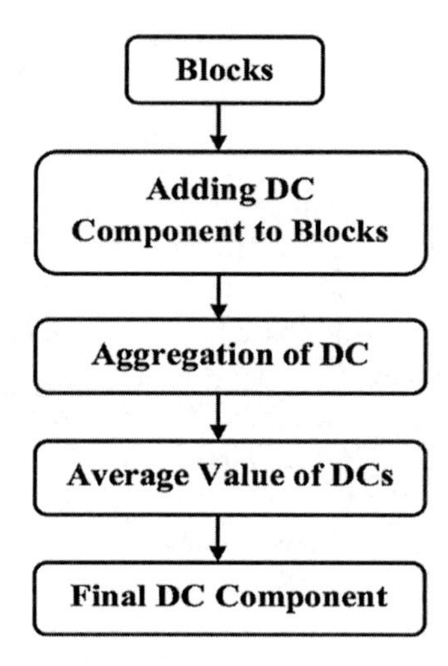

FIGURE 8.4 DC prediction technique.

8.3.3 IWT-LHT-Huffman Encoding

IWT-LHT-Huffman coding is an entropy encoding algorithm used for lossless data compression. Huffman coding is based on the incidence of occurrence of a pixel in images. The lower number of bits to encode the data occurs more frequently. For each image, the code is generated and stored in the codebook and the encoded data is used for decoding the image [49–52]. The coding efficiency is expressed as,

$$\eta = \frac{H(z)}{J(z)} \tag{8.14}$$

where $H(z) = -\sum_{d=1}^{r} P(c_d)\log\left[P(c_d)\right]$ and $J(z) = \sum_{d=1}^{r} J(c_d)log\left[P(c_d)\right]$.

8.3.4 IWT-LHT-Arithmetic Encoding

IWT-LHT-Arithmetic encoding is a lossless compression technique. In this, it stores frequently used characters with less number of bits and non-frequently used characters with more number of bits [51,53]. The input string P comprises of Q letters and q unique symbols. After the completion of the sorting operation of weights, m is considered to be the index of the symbol when each unique symbol is symbolized by a_m. Let us consider that every symbol takes place by k_m times, the number of occurrences is expressed as,

$$w_m = \frac{k_m}{q} \tag{8.15}$$

The length is expressed as $-\log_2(w_m)$. Hence the compressed length is $\sum_{m=1}^{q} -k_m \log_2(w_m)$.

8.3.5 Algorithm

The algorithm is as follows:

Step 1. Given the input medical image MRI.
Step 2. To separate LL, HL, LH, and HH sub-bands from the given input image decompose it by 2-D LIWT.
Step 3. Justify non-LL sub-bands based on the threshold.
Step 4. From the obtained results, encode the coefficients directly or truncation is performed after applying 2-D LHT techniques and truncate.
Step 5. Apply 2-D LHT only for LL band and DC can be predicted by DCP and truncate.
Step 6. By IWT-LHT-Huffman encoder/IWT-LHT-Arithmetic encoder, encode the resultant DC and AC coefficients separately.
Step 7. Finally, analyze MSE, PSNR, Normalized Correlation (NK), structural content (SC), Laplacian mean square error (LMSE), structural similarity index metric (SSIM), quality index (Q), and CR.

8.4 RESULTS AND DISCUSSION

Experiments were done to check out the performance of IWT-LHT-Huffman, IWT-LHT-Arithmetic lossless, and JPEG lossless medical image compression system which is executed on a laptop having the configuration of 2.3 GHz Intel Core i3 processor with 4 GB of RAM. The algorithms are implemented by using MATLAB® software version 2016a. In the medical field, modality is a method of therapeutic approach or a method of diagnosis. Radiography, ultrasound, CT, and MRI are the various types of modalities that are used to acquire images of the body in the field of medical imaging. In this research work, the DICOM image is taken as input which is a brain protocol magnetic resonance image. The size and bits per pixel of these images are 128×128 and 12 respectively. Input images are depicted in Figure 8.5.

These input images undergo both wavelet and Hadamard transform. First, the input images are separated into sub-bands. Based on the sub-bands, lossless Hadamard transformation was applied to the LL region to discard correlation within the block and DCP to eliminate correlation among the adjacent blocks. Non-LL sub-bands are authenticated as the NTB and encoding can be done either directly or after truncation. Finally, the resultant coefficient is encoded by IWT-LHT-Huffman encoder or IWT-LHT-Arithmetic encoder. The output images for IWT-LHT-Huffman compression are shown in Figures 8.6 and 8.7.

The output images for IWT-LHT-Arithmetic compression are depicted in Figures 8.8 and 8.9. The output images for JPEG lossless compression are depicted in Figure 8.10.

The proposed compression approach IWT-LHT-Arithmetic coder was compared with the IWT-LHT-Huffman coder and JPEG lossless algorithm, and it is validated by the following metrics. In the domain of image compression, if an image is reconstructed from the distorted image, its quality has to be estimated quantitatively. MSE, PSNR, normalized cross-correlation, SC, LMSE, SSIM, Q, and CR are the quality measures used for validating the efficacy of various compression approaches.

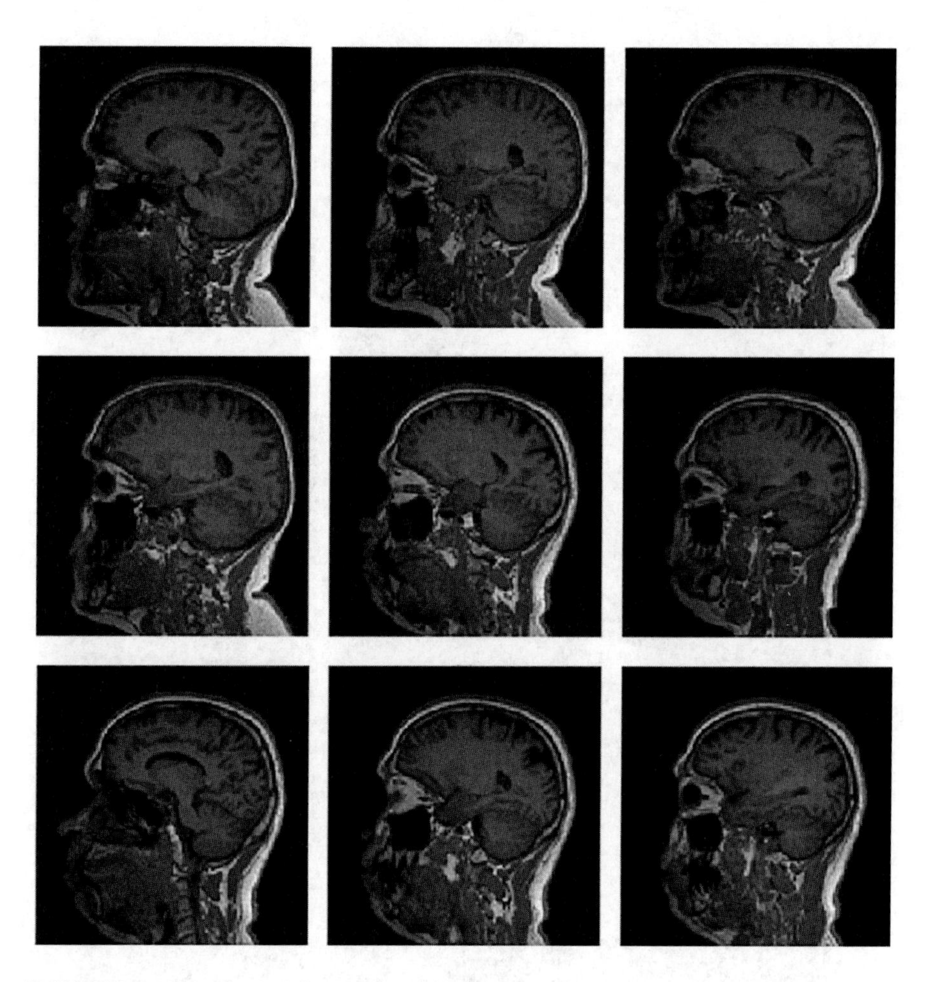

FIGURE 8.5 Input images for compression algorithm (ID1–ID9).

Assume the size of the image as $m \times n$. In addition, let us consider $I_o(m,n)$ and $I_r(m,n)$ be the original and reconstructed images, respectively. MSE and PSNR can be determined as follows [54]:

$$\text{MSE} = \frac{1}{mn}\sum_{m=1}^{i}\sum_{n=1}^{j}\left[I_o(m,n) - I_r(m,n)\right]^2 \tag{8.16}$$

$$\text{PSNR} = 10 \log \frac{f_{\max}^2}{\text{MSE}} \tag{8.17}$$

where $f_{\max}$ corresponds to the maximum value of pixels in an image. The smallest values of MSE and the largest values of PSNR lead to better efficiency. The plot of

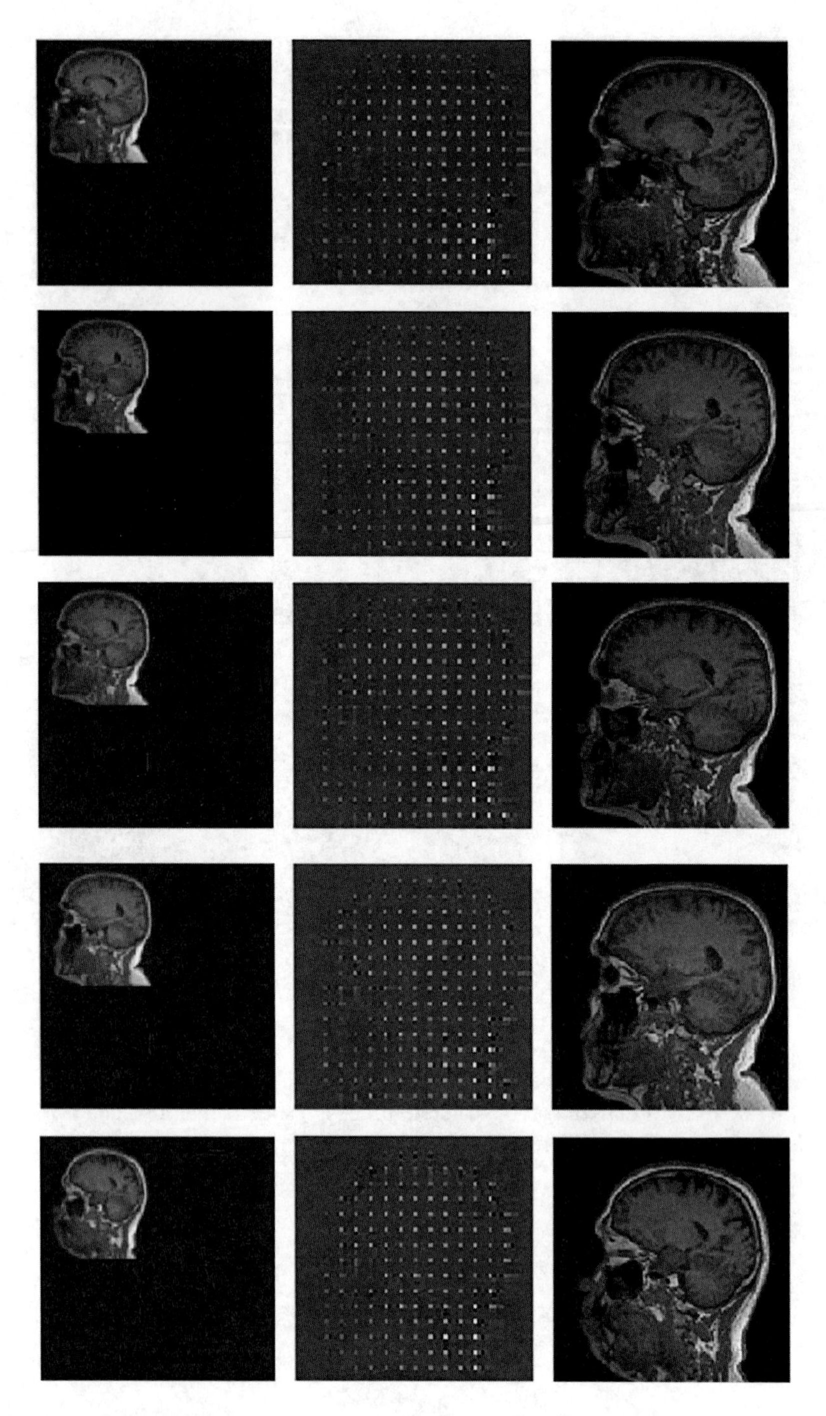

FIGURE 8.6 Integer wavelet transform (IWT)-lossless Hadamard transform (LHT)-Huffman compression for input images of ID1–ID5 in each row.

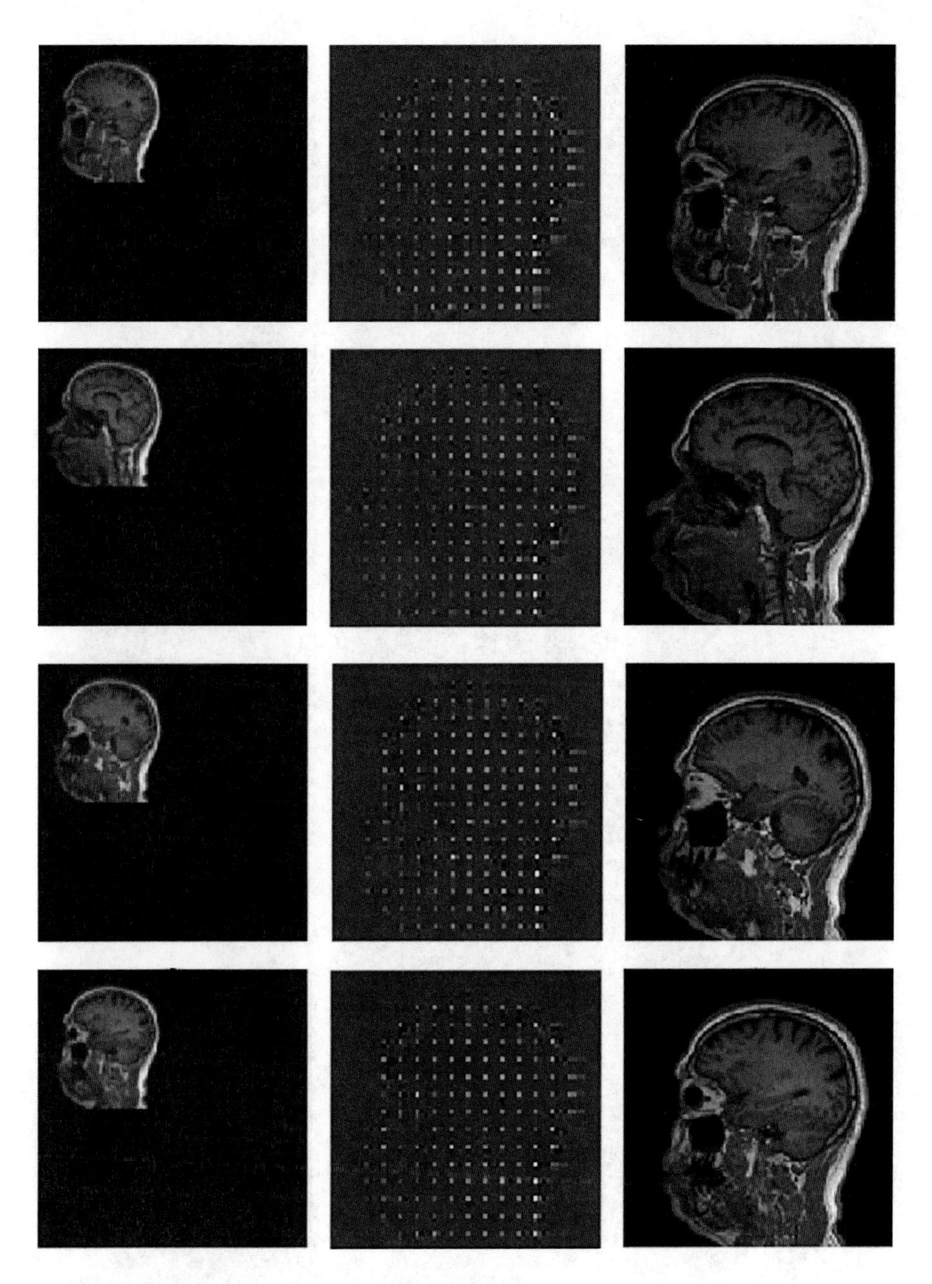

FIGURE 8.7 Integer wavelet transform (IWT)-lossless Hadamard transform (LHT)-Huffman compression for input images of ID6–ID9 in each row.

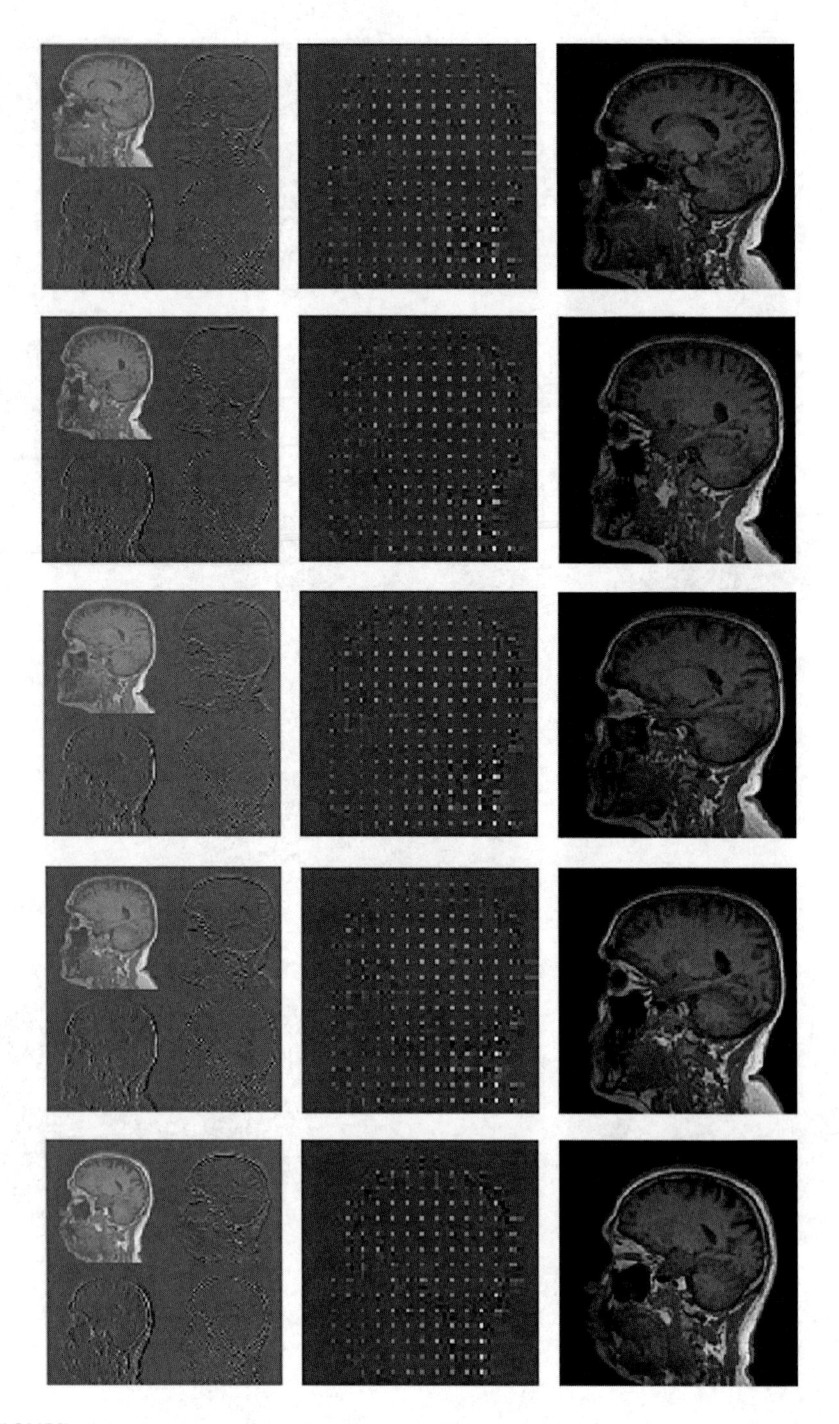

FIGURE 8.8 Integer wavelet transform (IWT)-lossless Hadamard transform (LHT)-Arithmetic compression for input images of ID1–ID5 in each row.

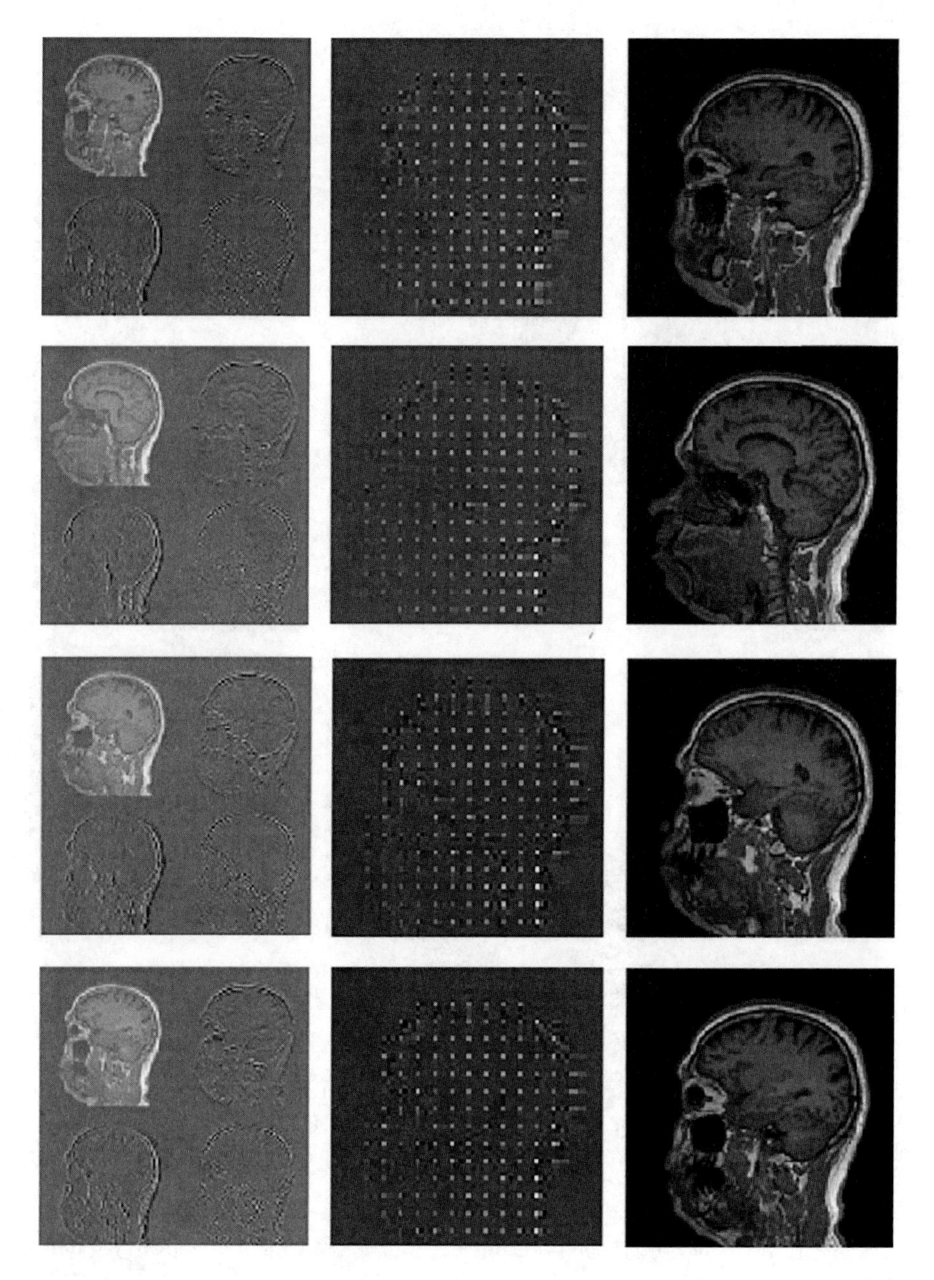

FIGURE 8.9 Integer wavelet transform (IWT)-lossless Hadamard transform (LHT)-Arithmetic compression for input images of ID6–ID9 in each row.

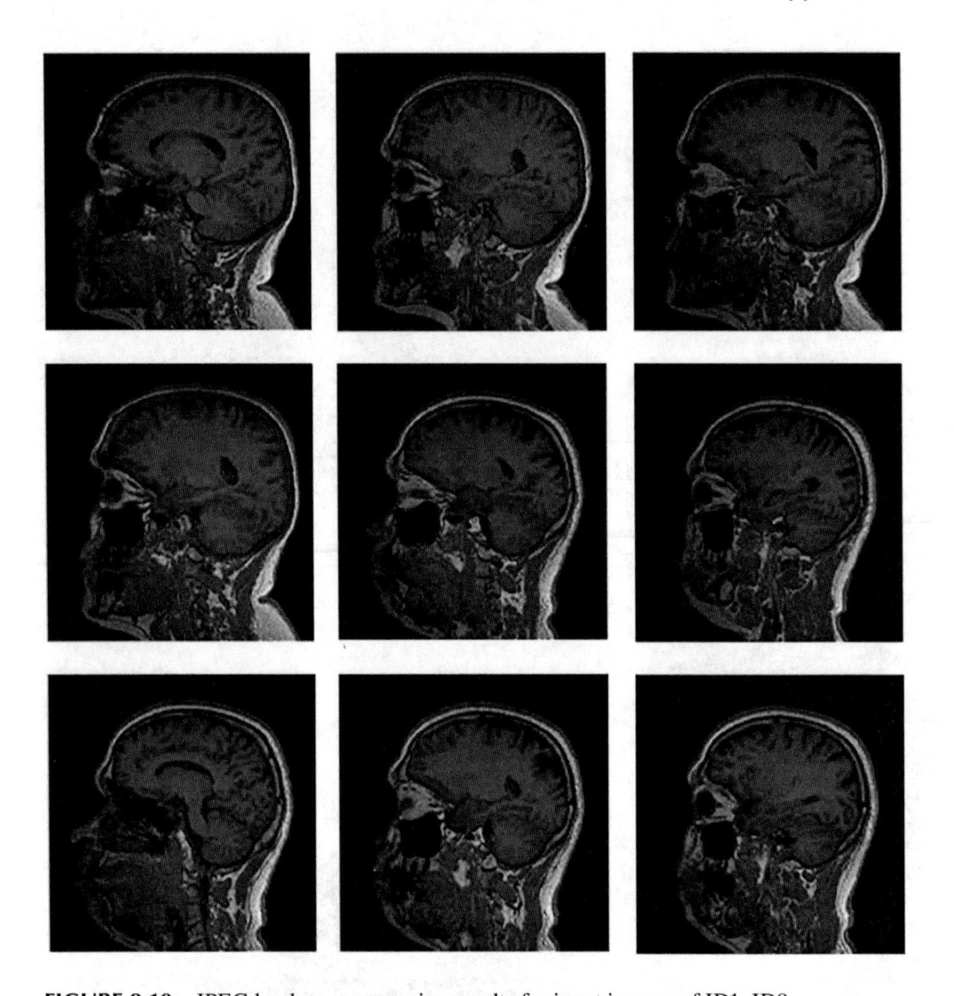

FIGURE 8.10 JPEG lossless compression results for input images of ID1–ID9.

MSE and PSNR is delineated in Figures 8.11 and 8.12, respectively. The hybrid compression technique with arithmetic coder yields superior performance when compared with the hybrid compression of Huffman coder and JPEG lossless compression techniques.

NK and SC point out the similitude between the original and reconstructed images. If the values of NK and SC either approaches 1 or low it tends to better the image quality.

$$\text{NK} = \frac{\sum_{m=1}^{i}\sum_{n=1}^{j}\left[I_o(m,n) - I_r(m,n)\right]}{\sum_{m=1}^{i}\sum_{n=1}^{j}\left[I_o(m,n)\right]^2} \tag{8.18}$$

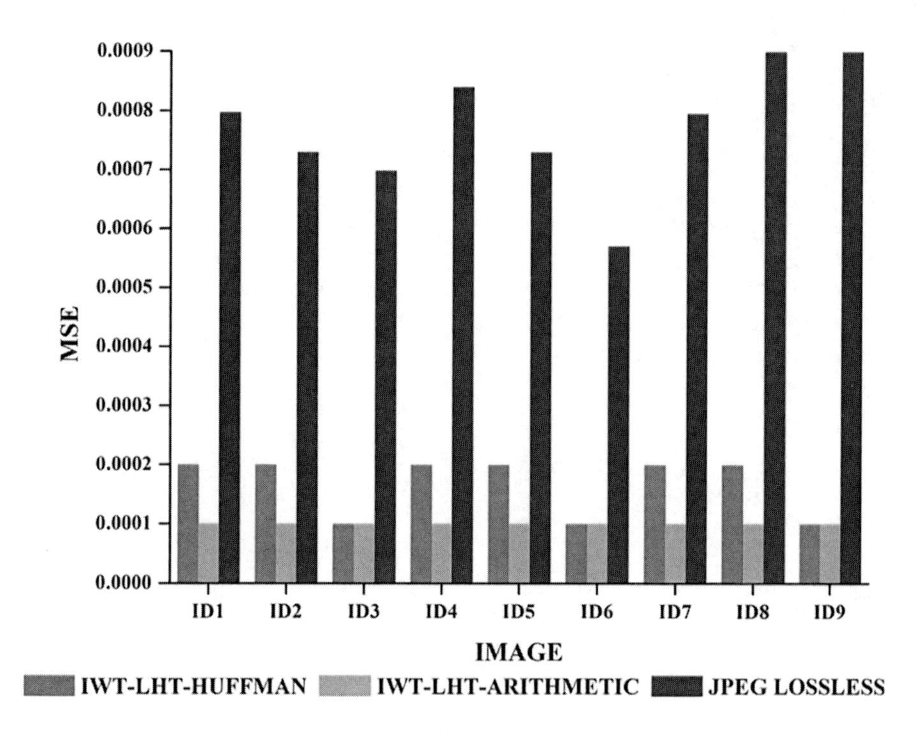

FIGURE 8.11 Mean square error (MSE) plot of compression algorithms.

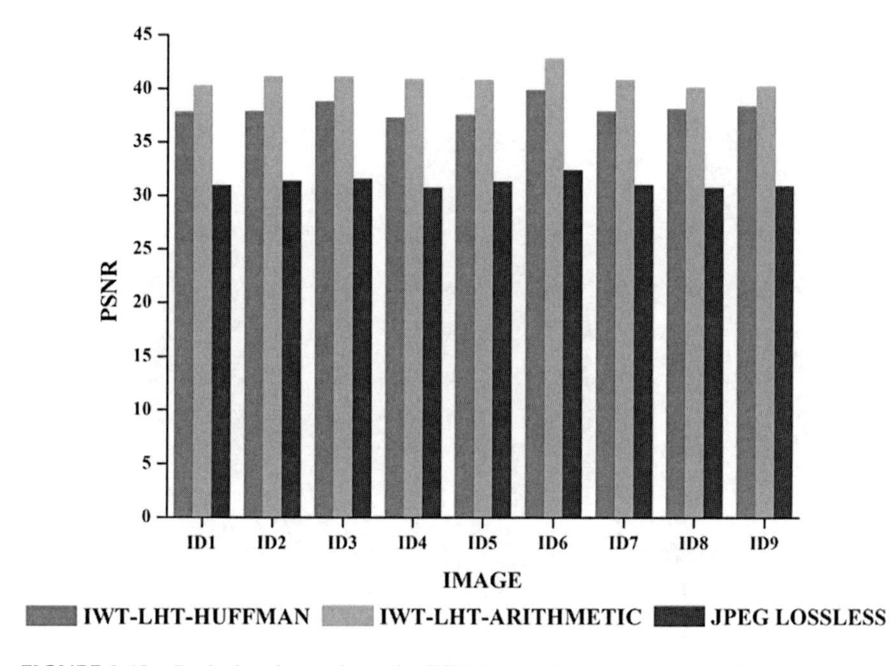

FIGURE 8.12 Peak signal-to-noise ratio (PSNR) plot of compression algorithms.

$$\text{SC} = \frac{\sum_{m=1}^{i}\sum_{n=1}^{j}\left[I_o(m,n)\right]^2}{\sum_{m=1}^{i}\sum_{n=1}^{j}\left[I_r(m,n)\right]^2} \tag{8.19}$$

NK and SC plot is depicted in Figures 8.13 and 8.14, respectively. It also reveals the superiority of the hybrid compression technique with the arithmetic coder.

LMSE plot is shown in Figure 8.15 which uses Laplacian kernel and it is elicited from the edge measurement. The value of LMSE should be as small as possible to ensure better image quality [55].

$$\text{LMSE} = \frac{\sum_{m=1}^{i}\sum_{n=1}^{j}\left\{L\left[I_o(m,n)\right]-L\left[I_r(m,n)\right]\right\}^2}{\sum_{m=1}^{i}\sum_{n=1}^{j}\left\{L\left[I_o(m,n)\right]\right\}^2} \tag{8.20}$$

The hybrid compression technique with arithmetic coder is efficient with regard to LMSE.

The Q index [56] and SSIM [57] displayed in Figures 8.16 and 8.17 can be expressed as follows.

$$Q = \frac{\sigma_{mn}}{\sigma_m\sigma_n} \times \frac{2\overline{xy}}{(\overline{x})^2+(\overline{y})^2} \times \frac{2\sigma_m\sigma_n}{\sigma_m{}^2+\sigma_n{}^2} \quad Q = \frac{\sigma_{mn}}{\sigma_m\sigma_n} \times \frac{2\overline{xy}}{(\overline{x})^2+(\overline{y})^2} \times \frac{2\sigma_m\sigma_n}{\sigma_m{}^2+\sigma_n{}^2} \tag{8.21}$$

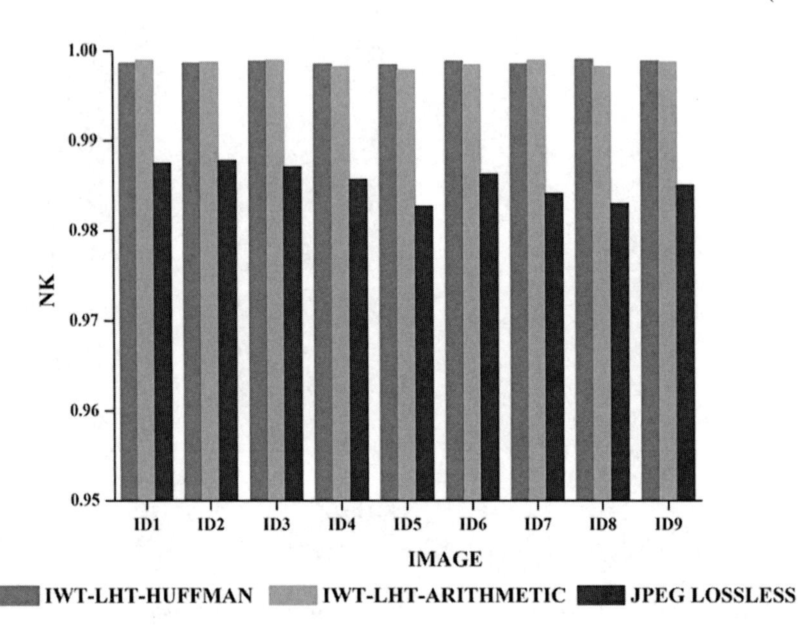

FIGURE 8.13 NK plot of compression algorithms.

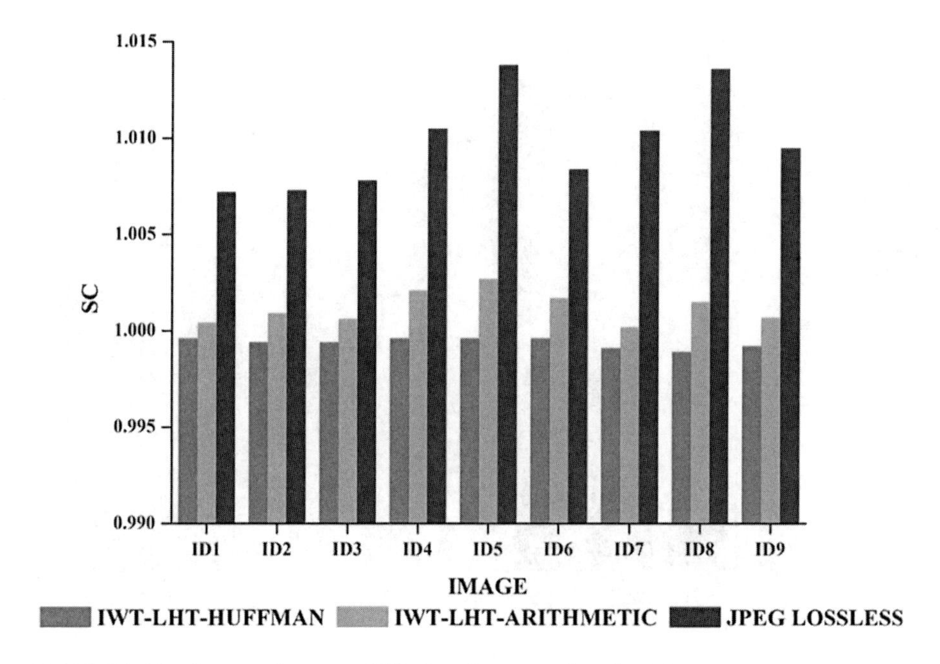

FIGURE 8.14 Structural content (SC) plot of compression algorithms.

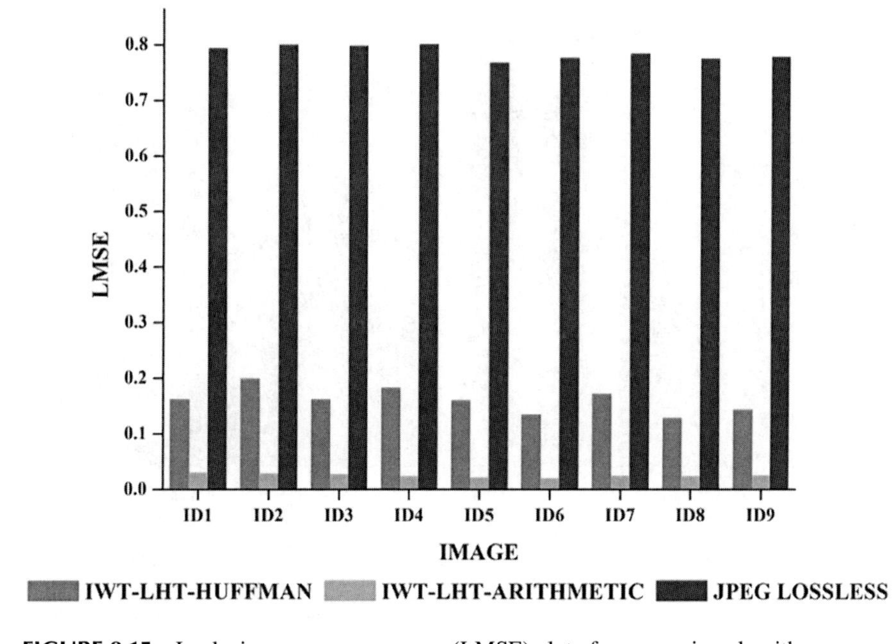

FIGURE 8.15 Laplacian mean square error (LMSE) plot of compression algorithms.

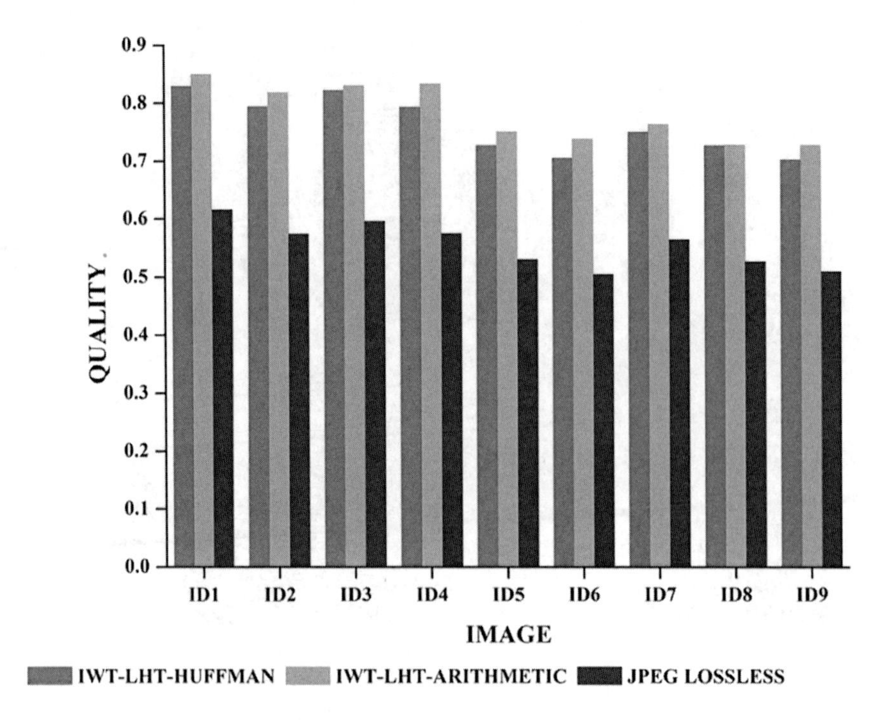

FIGURE 8.16 Quality index (Q) plot of compression algorithms.

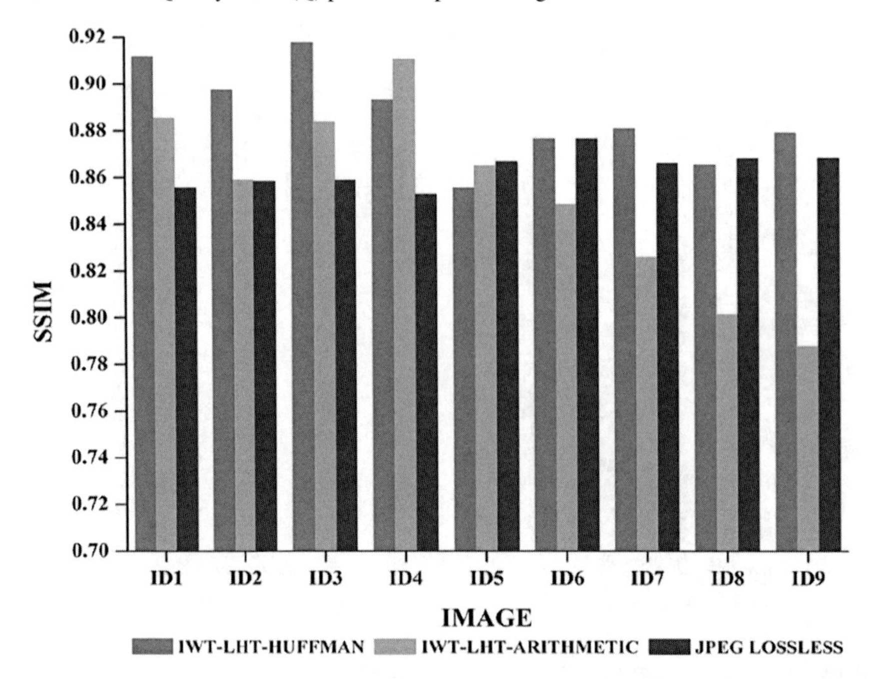

FIGURE 8.17 Structural similarity index metric (SSIM) plot of compression algorithms.

$$\text{SSIM} = \frac{(2\mu_m\mu_n + \alpha_1)(2\sigma_{mn} + \alpha_2)}{(\mu_m^2 + \mu_n^2 + \alpha_1)(\sigma_m^2 + \sigma_n^2 + \alpha_2)} \tag{8.22}$$

where μ_m and μ_n represent the mean, σ_m and σ_n symbolize the standard deviation, σ_{mn} indicates correlation coefficient, α_1 and α_2 are constants. The hybrid compression technique with arithmetic coder is proficient in terms of Q index and SSIM.

The proportion between the size of the original and reconstructed image is known as CR.

$$\text{CR} = \frac{\text{Uncompressed file size}}{\text{Compressed file size}} \tag{8.23}$$

The compression algorithm performance is also validated by the CR which is depicted in Tables 8.1 and 8.2. The CR is high for the IWT-LHT-Huffman algorithm when compared with IWT-LHT-Arithmetic and JPEG lossless compression techniques. The computation time is calculated which is depicted in Table 8.3 for analyzing the complexity of the algorithm and it shows IWT-LHT-Arithmetic yields reduced complexity than the other compression algorithms. The teleradiology refers to the storage and transfer of medical data and is a part of telemedicine. A huge volume of medical data are generated and an efficient compression technique is required and the outcome of this work is an aid for teleradiology application.

TABLE 8.1
Performance Evaluation for Image Memory Size

			Compressed Image Memory Size		
Sl. No.	**Image Name**	**Uncompressed Image Memory Size (bytes)**	**IWT-LHT-Huffman (bytes)**	**IWT-LHT-Arithmetic (bytes)**	**JPEG Lossless (bytes)**
1	ID1	24576	5590.75	8021.5	2097152
2	ID2	24576	5282.38	7870.38	2097152
3	ID3	24576	5605.88	7607.38	2097152
4	ID4	24576	5252.63	7481.88	2097152
5	ID5	24576	5396.13	7317.13	2097152
6	ID6	24576	5348.25	7549.5	2097152
7	ID7	24576	5376.88	7741.38	2097152
8	ID8	24576	5309.38	7315.75	2097152
9	ID9	24576	5313.75	7700.63	2097152

TABLE 8.2
Performance Evaluation for Compression Ratio

Sl. No.	Image Name	IWT-LHT-Huffman	IWT-LHT-Arithmetic	JPEG Lossless
1	ID1	4.4	3.06	0.05
2	ID2	4.65	3.12	0.05
3	ID3	4.38	3.23	0.05
4	ID4	4.68	3.28	0.05
5	ID5	4.55	3.36	0.05
6	ID6	4.6	3.26	0.05
7	ID7	4.57	3.17	0.05
8	ID8	4.63	3.36	0.05
9	ID9	4.62	3.19	0.05

TABLE 8.3
Performance Evaluation for Computation Time in Seconds

Sl. No.	Image Name	IWT-LHT-Huffman	IWT-LHT-Arithmetic	JPEG Lossless
1	ID1	0.552277	0.812879	1.058826
2	ID2	0.815635	0.394277	1.056524
3	ID3	2.791036	0.439501	1.068123
4	ID4	1.285199	0.503316	1.041921
5	ID5	0.762111	0.360682	1.053639
6	ID6	0.642956	0.358993	1.023719
7	ID7	0.884054	0.374892	1.10726
8	ID8	0.731911	0.381406	0.830658
9	ID9	1.142515	0.394111	1.129071

8.5 CONCLUSION

In this research work, medical images are compressed by using a block-based compression technique. The IWT-LHT-Arithmetic coder generates superior results when compared with IWT-LHT-Huffman and JPEG lossless compression models. To achieve this, simple validation of NTB, appropriate truncation, and DCP are needed. The algorithms used in this work are tested with nine MRI medical images and it provides a better result for the diagnosis. Performance is computed based on the metrics: MSE, PSNR, NK, SC, LMSE, Q index, SSIM, and CR. Hence IWT-LHT-Huffman and IWT-LHT-Arithmetic coding algorithms will effectively compress the medical image for efficiently transferring and storage for teleradiology applications. If parallel processing is used, the computational time gets reduced and hence in the future, this work is extended to improve the performance by the implementation of IWT-LHT-Arithmetic encoding.

REFERENCES

1. Sumalatha R, Subramanyam MV. Hierarchical lossless image compression for telemedicine applications. *Procedia Computer Science.* 2015;54:838–848.
2. Taubman D. High performance scalable image compression with EBCOT. *IEEE Transactions on Image Processing.* 2000;9:1158–1170.
3. Carpentieri B, Weinberger MJ, and Seroussi G. Lossless compression of continuous-tone images. *Proceedings of the IEEE.* 2000;88(11):1797–1809.
4. Weinberger MJ, Seroussi G, and Sapiro G. The LOCO-I lossless image compression algorithm: Principles and standardization into JPEG-LS. *IEEE Transactions on Image Processing.* 2000;9(8):1309–1324.
5. Yng TLB, Lee BG, and Yoo H. Low complexity, lossless frame memory compression using modified Hadamard transform and adaptive Golomb-Rice coding. *IADIS Int. Conf. Comput. Gr. Vis.*, pp. 89–96, 2008.
6. Philips W., and Denecker K. "A lossless version of the Hadamard transform," in: *Proc. Pro RISC Workshop Circuits Syst. Signal Process*, 1997, pp. 443–450.
7. Hussain AJ, Al-Fayadh A, and Radi N. Image compression techniques: A survey in lossless and lossy algorithms. *Neurocomputing.* 2018;300:44–69.
8. Mochizuki T. Bit pattern redundancy removal for Hadamard transformation coefficients and its application to lossless image coding. *Electronics and Communications in Japan.* 1997;80(6):1–10.
9. Tan KCB, and Arslan T. Low power embedded extension algorithm for lifting based discrete wavelet transform in JPEG2000. *Electronics Letters.* 2001;37(25):1328–1330.
10. Li Q., Ren G., Wu Q., and Zhang X. Rate pre-allocated compression for mapping image based on wavelet and rate-distortion. *International Journal for Light and Electron Optics.* 2013;124(14):1836–1840.
11. Lin C, Zhang B, and Zheng YF. Packed integer wavelet transform constructed by lifting scheme. *IEEE Transactions on Circuits and Systems for Video Technology.* 2000;10(8):1496–1150.
12. Al-Sulaifanie AK, Ahmadi A, and Zwolinski M. Very large scale integration architecture for integer wavelet transform. *IET Computers & Digital Techniques.* 2010;4(6):471–483.
13. Zhang M, and Tong X. Joint image encryption and compression scheme based on IWT and SPIHT. *Optics and Lasers in Engineering.* 2017;90:254–274.
14. Grangetto M, Magli E, Martina M, and Olmo G. Optimization and implementation of the integer wavelet transform for image coding. *IEEE Transactions on Signal Processing.* 2002;11:596–604.
15. Alotaibi RA, and Elrefaei LA. Text-image watermarking based on integer wavelet transform (IWT) and discrete cosine transform (DCT). *Applied Computing and Informatics.* 2019;15(2):191–202.
16. Yang L, He X, Zhang G, Qinga L, and Che T. A low complexity block-based adaptive lossless image compression. *Optik* 2013;124:6545–6552.
17. Al-Khafaji G, and George LE. Fast lossless compression of medical images based on polynomial. *International Journal of Computer Applications.* 2013;70(15):0975–8887.
18. Rajkumar TMP, and Latte M. An efficient ROI encoding based on LSK and fractal image compression. *The International Arab Journal of Information Technology.* 2015;12(3):220–228.

19. Al-Khafaji G. Wavelet transform and polynomial approximation model for lossless medical image compression. *International Journal of Advanced Research in Computer Science and Software Engineering.* 2014;4(3):584–587.
20. Sundaresan M, and Devika E. Efficient integrated coding for compound image compression. *International Journal of Computational Intelligence and Informatics.* 2012;2(2):153–158.
21. Bruckmann A, and Uhl A. Selective medical image compression using wavelet techniques. *Journal of Computing and Information Technology.* 1998;2:203–213.
22. Chandraraju T, and Radhakrishnan S. Image encoder architecture design using dual scan based DWT with vector quantization. *Materials Today: Proceedings.* 2018;5(1):572–577.
23. Elsayed HA, Ghazi O, and Guirguis S. Volumetric medical images lossy compression using stationary wavelet transform and linde-buzo-gray vector quantization. *Research Journal of Applied Sciences, Engineering and Technology.* 2017;14(9):352–360.
24. Hou X, Han M, Gong C, and Qian X. SAR complex image data compression based on quadtree and zerotree coding in discrete wavelet transform domain: A comparative study. *Neurocomputing.* 2015;148:561–568.
25. Nowaková J, Prílepok M, Snášel V. Medical image retrieval using vector quantization and fuzzy S-tree. *Journal of Medical Systems.* 2017;41(18):1–18.
26. Phanprasit T. "Compression of medical image using vector quantization," *IEEE Conference on Biomedical Engineering*, Thailand, October 6, pp. 1–4. 2013.
27. Hosseini SM, and Naghsh-Nilchi AR. Medical ultrasound image compression using contextual vector quantization. *Computers in Biology and Medicine.* 2012;42(7):743–750.
28. Chen YT, and Tseng DC. Wavelet-based medical image compression with adaptive prediction. *Computerized Medical Imaging and Graphics.* 2007;31(1):1–8.
29. Ayoobkhan MU, Chikkannan E, and Ramakrishnan K. Lossy image compression based on prediction error and vector 219 quantisation. *EURASIP Journal on Image and Video Processing.* 2017;2017(35):1–13.
30. Kuo CH, Chou TC, Wang TS. An efficient spatial prediction based image compression scheme. *IEEE Transactions on Circuits and Systems for Video Technology.* 2002;12(10): 850–856.
31. Sun M, He X, Xiong S, Ren C, and Li X. Reduction of JPEG compression artifacts based on DCT coefficients prediction. *Neurocomputing.* 2020;384: 335–345.
32. Tiwari M. STW and SPIHT wavelet compression using MATLAB wavelet tool for color image. *arXiv* preprint arXiv:2002.08897.
33. Jain A, and Datar A. Spatial video compression using EZW 3D-SPIHT WDR & ASWDR techniques. *International Journal of Advanced Research in Computer Science and Software Engineering.* 2013;3(7): 1413–1420.
34. Minnen D., Ballé J., and Toderici G.D. Joint autoregressive and hierarchical priors for learned image compression. In *Advances in Neural Information Processing Systems* 2018 (pp. 10771–10780).
35. Ayzik S, Avidan S. Deep image compression using decoder side information. *arXiv* preprint arXiv:2001.04753.
36. Cheng Z, Sun H, Takeuchi M, and Katto J. Deep convolutional autoencoder-based lossy image compression. In *2018 Picture Coding Symposium (PCS)*, 2018 (pp. 253–257). IEEE.
37. Wani N., and Raza K. *Multiple Kernel Learning Approach for Medical Image Analysis. Soft Computing Based Medical Image Analysis*, Elsevier, 31–47, 2018.
38. Raza K, and Singh NK. A tour of unsupervised deep learning for medical image analysis. *arXiv* preprint arXiv:1812.07715. 2018.
39. Zhang S, Yao L, Sun A, and Tay Y. Deep learning based recommender system: A survey and new perspectives. *ACM Computing Surveys (CSUR).* 2019;52(1):1–38.

40. Kumar SN, Fred AL, and Varghese PS. Compression of CT images using contextual vector quantization with simulated annealing for telemedicine application. *Journal of Medical Systems.* 2018;42(11):218.
41. Kumar SN, Fred AL, Kumar HA, and Varghese PS. Lossless compression of CT images by an improved prediction scheme using least square algorithm. *Circuits, Systems, and Signal Processing.* 2020;39(2): 522–542.
42. Boix M, and Canto B. Wavelet Transform application to the compression of images. *Mathematical and Computer Modelling.* 2010;52(7–8):1265–1270.
43. Al-Fayadh A, and Abdulkareem M. Improved transform based image compression methods. *Applied Mathematical Sciences.* 2017;11(47):2305–2314.
44. Bruni V, Cotronei M, and Pitolli F. A family of level-dependent biorthogonal wavelet filters for image compression. *Journal of Computational and Applied Mathematics.* 2020;367:112467.
45. Kim H, and Li CC. Unified image compression using reversible and fast biorthogonal wavelet transform. In *Proceedings IWISP'96*, 1996 (pp. 263–266). Elsevier Science Ltd.
46. Ghrare SE, and Khobaiz AR. Digital image compression using block truncation coding and Walsh Hadamard transform hybrid technique. In *2014 International Conference on Computer, Communications, and Control Technology (I4CT)*, 2014 (pp. 477–480). IEEE.
47. Parfieniuk M. Lifting-based algorithms for computing the 4-point Walsh-Hadamard transform. In *2019 Signal Processing Symposium (SPSympo)*, 2019 (pp. 221–226). IEEE.
48. Venugopal D, Mohan S, and Raja S. An efficient block based lossless compression of medical images. *Optik.* 2016;127(2):754–758.
49. Dubey VG, Singh J. 3D medical image compression using Huffman encoding technique. *International Journal of Scientific and Research Publications.* 2012;2(9): 1–3.
50. Kumar T., and Kumar R. Medical image compression using hybrid techniques of DWT, DCT and Huffman coding. *International Journal of Innovative Research in Electrical, Electronics, Instrumentation and Control Engineering.* 2015;3(2):54–60.
51. Howard PG, and Vitter JS. Parallel lossless image compression using Huffman and arithmetic coding. In *Data Compression Conference*, 1992. (pp. 299–308). IEEE.
52. Yuan S, and Hu J. Research on image compression technology based on Huffman coding. *Journal of Visual Communication and Image Representation.* 2019;59:33–38.
53. Wu J, Xu Z, Jeon G, Zhang X, and Jiao L. Arithmetic coding for image compression with adaptive weight-context classification. *Signal Processing: Image Communication.* 2013;28(7):727–735.
54. Fred AL, Kumar SN, Kumar HA, and Abisha W. Bat optimization based vector quantization algorithm for medical image compression. In *Nature Inspired Optimization Techniques for Image Processing Applications*, 2019 (pp. 29–54). Springer.
55. Kumar SN, Fred AL, Kumar HA., and Varghese PS. Lossless compression of CT images by an improved prediction scheme using least square algorithm. *Circuits, Systems, and Signal Processing.* 2020;39(2):522–542.
56. Wang Z, and Bovik AC. A universal image quality index. *IEEE Signal Processing Letters.* 2002;9(3):81–84.
57. Li C, and Bovik AC. Content-partitioned structural similarity index for image quality assessment. *Signal Processing: Image Communication.* 2010;25(7):517–526.

9 Improved FCM Based on Gaussian Kernel and Crow Search Optimization for ROI Extraction on Corona Virus Disease (COVID-19) CT Images

S. N. Kumar
Amal Jyothi College of Engineering

Lenin Fred A., L. R. Jonisha Miriam and Ajay Kumar H.
Mar Ephraem College of Engineering and Technology

Parasuraman Padmanabhan and Balázs Gulyás
Cognitive Neuroimaging Centre (CONIC), Lee Kong Chian School of Medicine, Nanyang Technological University

CONTENTS

9.1 INTRODUCTION

Segmentation is the technique of isolating the desired region of interest (ROI) from an image and perspective of medical images; ROI depicts the anatomical organ or anomalies like a tumor, cyst, etc. A wide variety of segmentation algorithms are there for medical images and some of the widely used segmentation algorithms are discussed in [1,2]. The author proposed a modification in the fuzzy c-means (FCM) by normalized Canberra distance by eliminating the Euclidean distance that effectively removes the noise and helps in the accurate segmentation of the brain lesions. The method performs better than FCM, spatial FCM, kernelized FCM (KFCM), and adaptively regularized kernel-based FCM (ARKFCM) algorithms in terms of Dice, Jaccard, precision, and recall metrics [3]. For the segmentation of the nonuniform background of the CT images, the author proposed an adaptive scale-kernel fuzzy clustering approach that relies on the two-dimensional histogram between the clustered pixel and its neighborhood mean which is iterated by Lagrange multiplier. The performance of brain segmentation was analyzed qualitatively and quantitatively and compared with other clustering algorithms [4].

The author introduced a framework with average and median filter followed by Gaussian radial basis kernel function and clustering of the brain tissue was done by ARKFCM. This method has high segmentation accuracy when validated and compared with Gaussian kernel-based FCM with average filter (GKFCM1), GKFCM with median filter (GKFCM2), fuzzy local information c-means (FLICM), kernel-weighted FLICM, multiplicative intrinsic component optimization (MICO), adaptively regularized kernel-based FCM with average filter (ARKFCM1), ARKFCM with median filter (ARKFCM2), and ARKFCM with weighted image soft clustering approaches [5]. The author suggested a hybrid method for the brain image clustering using the farthest point first algorithm and fuzzy clustering algorithm. This method effectively segments WM, GM, and CSF of the MR brain images. The proposed method outperforms with 91% Jaccard similarity when compared with FCM, KFCM, ARKFCM, and farthest point first techniques [6]. The MR brain images segmentation of WM, GM, and CSF was done by the improved ARKFCM with spatial constraints and the proposed technique was compared with other three variants of FCM and it achieves higher Jaccard similarity and accuracy in Brain Web database images [7]. The adaptive kernel-based FCM clustering algorithm with spatial constraints was proposed that considers the neighborhood pixel in the clustering. The proposed method was tested with synthetic images and performance was better, even the noise level greater than 3% and tested on real-time medical images also [8].

A modified robust fuzzy c-means technique was put forward for the ROI delineation on MR brain images. The segmentation accuracy of the algorithm was tested and validated at different noise levels in synthetic and medical images with the Dice coefficient and also compared with FCM, enhanced fuzzy c-means, fast generalized FCM, MICO, and ARKFCM algorithms [9]. The author suggested a pixel-based FCM image segmentation algorithm that uses a bilateral filter as a kernel to form a pixel image that avoids over-segmentation [10]. For the segmentation of brain tissues GM, WM, and CSF, the author proposed robust kernelized local information fuzzy c-means that uses Gaussian radial basis function as distance metrics. The segmentation algorithm was tested on brain MR images and achieves better accuracy and Jaccard similarity when corrupted with Gaussian and Rician noise. The computation time is also much low when compared with other clustering techniques [11]. The fuzzy possibilistic c-means algorithm was found to be effective in the segmentation of lungs CT images, when compared with the FCM and modified FCM algorithms [12]. ANN and fuzzy clustering were found to be effective in lung cancer detection [13]. The ant colony optimization was found to be efficient in the initialization of cluster centroids in the FCM algorithm and produces superior results, in contrast with the classical FCM algorithm on medical images [14]. A novel modified fast adaptive FCM was proposed for the segmentation of lung nodules [15]. The fuzzy logic when coupled with the gene regulatory network yields efficient results for the analysis of biological data [16]. A hybrid ROI extraction technique comprising of watershed and FCM was used for the medical images; the preprocessing was done by morphological operations and the results outperform the classical FCM algorithm [17]. The deep learning algorithms play a vital role in medical image analysis and give promising results when compared with the classical machine learning algorithms [18]. The marker-controlled watershed algorithm with particle swarm optimization (PSO) for optimum cluster centroid selection for FCM was found to be efficient for the detection of lung cancer [19]. This chapter organization is as follows: Section 9.2 describes the improved ARKFCM algorithm based on crow search optimization (CSO), Section 9.3 depicts the results and discussion, and the conclusion is drawn in Section 9.4.

9.2 IMPROVED ARKFCM BASED ON CROW SEARCH OPTIMIZATION

The ARKFCM generates superior clustering results when equated with the FCM and variants of FCM like FLICM and IFCM techniques. This research work proposes the ARKFCM-crow segmentation algorithm that generates efficient results when compared with the ARKFCM technique. The CSO was evolved from the intelligent behavior of crows and was found to be proficient in the solving of engineering problems [20]. The optimization-based clustering algorithm was found to be robust in ROI extraction [18,21,22]. The flow diagram of the proposed clustering algorithm is depicted in Figure 9.1.

9.2.1 CROW SEARCH OPTIMIZATION

The optimization technique, called crow search optimization (CSO), was developed from the inspiration of crows that live in flocks. The nature of a crow is to collect and hide the food in a unique place. Moreover, it was remembered by that crow after a few months. Also, it stays away from the unique place when it feels that a thief crow follows it to save the food.

In CSO, a group of crows is considered as population and an individual crow is said to be a solution or search agent. Let us consider crows in a group that are ordered as *P*. At iteration *I*, the position of the crow is calculated by,

$$y_g^i = \left[y_{g1}^i, y_{g2}^i, \ldots, y_{ga}^i\right] \tag{9.1}$$

where *a* symbolizes the problem dimension.

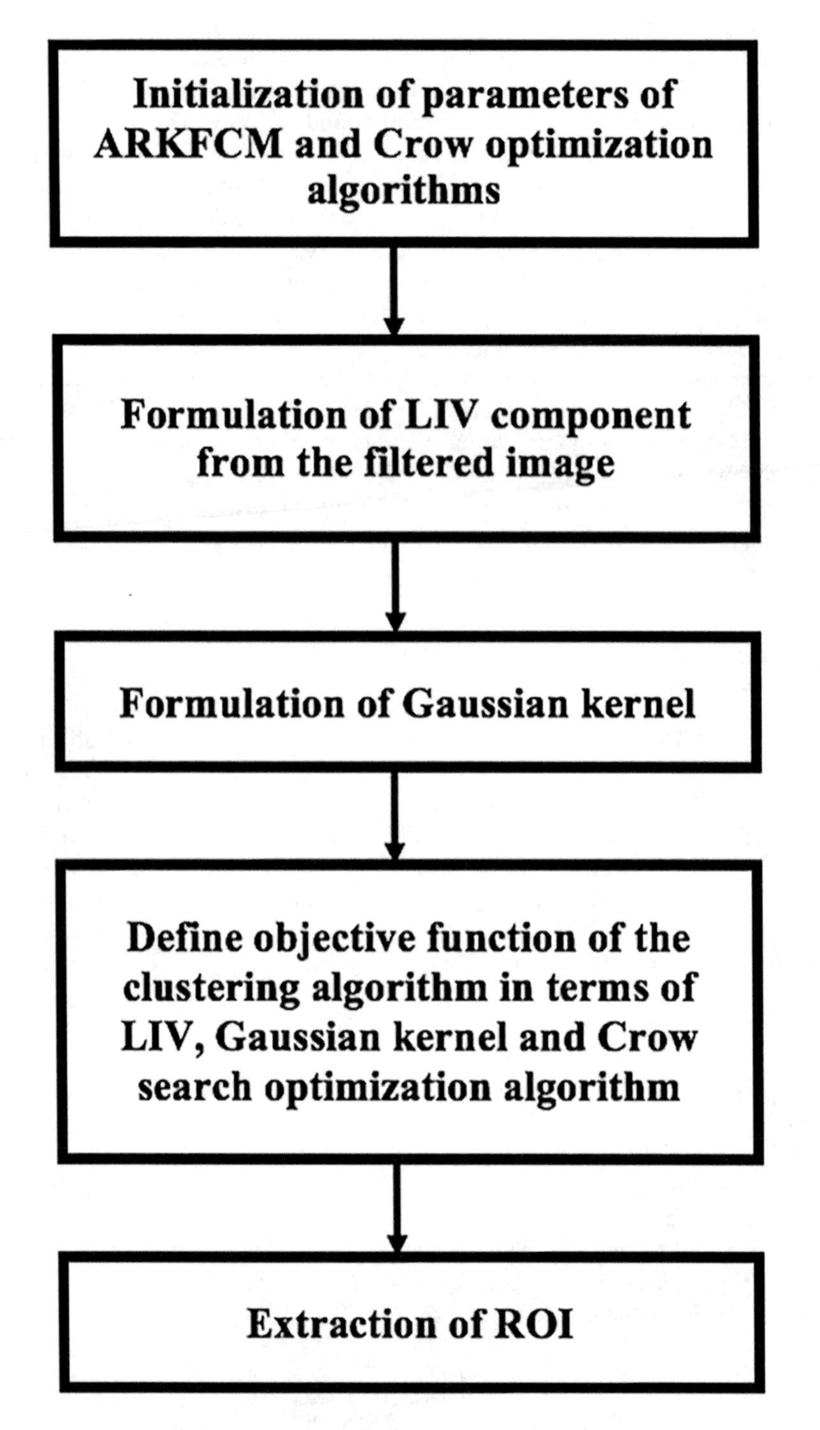

FIGURE 9.1 Flow diagram of the adaptively regularized kernel-based fuzzy c-means (ARKFCM) -crow segmentation algorithm.

The population is represented as

$$Y = \begin{bmatrix} y_{11}^{i} & y_{12}^{i} & \cdots & y_{1a}^{i} \\ y_{21}^{i} & y_{22}^{i} & \cdots & y_{2a}^{i} \\ \vdots & \vdots & \cdots & \vdots \\ y_{O1}^{i} & y_{O2}^{i} & \cdots & y_{O3}^{i} \end{bmatrix} \tag{9.2}$$

The memory of the population is given as,

$$Q = \begin{bmatrix} q_{11}^{i} & q_{12}^{i} & \cdots & q_{1a}^{i} \\ q_{21}^{i} & q_{22}^{i} & \cdots & q_{2a}^{i} \\ \vdots & \vdots & \cdots & \vdots \\ q_{O1}^{i} & q_{O2}^{i} & \cdots & q_{O3}^{i} \end{bmatrix} \tag{9.3}$$

Every solution is standardized by evaluating its fitness function $f\left(y_g^{(i+1)}\right)$. The position update takes place by anyone of the two states based on awareness probability. Let us consider the two crows to be g and h.

State 1: If the crow g follows crow h but crow h does not know that the crow g is following. At that time the crow g will find the food hiding place of crow h and crow g will update its position is as follows.

$$y_g^{(i+1)} = y_g^i + r_g \times fl_g^{(i)} \times \left(q_h^{(i)} - y_g^{(i)}\right) \tag{9.4}$$

where r_g denotes the random number lies in $[0,1]$ and $fl_g^{(i)}$ indicates flight length.

State 2: If crow h knows that the crow g is following. At the same time, crow h moves randomly to fool crow g.

$$y_g^{(i+1)} = \begin{cases} y_g^{(i)} + r_g \times fl_g^{(i)} \times \left(q_h^{(i)} - y_g^{(i)}\right), & r_h \geq AP_g^i \\ \text{Move to a random position}, & \text{Otherwise} \end{cases} \tag{9.5}$$

where AP_g^i denotes awareness probability.

The updation of memory is carried out by fitness value evaluation whose equation is expressed as follows.

$$q_g^{(i+1)} = \begin{cases} y_g^{(i+1)}, & \text{if } f\left(y_g^{(i+1)}\right) \text{ is better than } f\left(q_g^{(i)}\right) \\ q_g^i, & \text{Otherwise} \end{cases} \tag{9.6}$$

9.2.2 Improved Adaptive Regularized Kernel FCM with Crow Search Algorithm

The traditional FCM clustering is highly sensitive to artifacts and it yields less accuracy. To obtain better accuracy, filtering is used prior to ROI extraction; ARKFCM methodology uses regularization kernel factor. The median filter and average filter are used prior to segmentation. The regularization kernel from local intensity variance (LIV) is constructed. In the presence of noise, the value of LIV gets increased to attain measurable heterogeneity between the neighboring pixels. LIV is expressed as follows,

$$\mathrm{LIV}_n = \frac{\sum_{m \in R_n} \left(z_m - \overline{z_n}\right)^2}{R_s \times \left(\overline{z_n}\right)^2} \tag{9.7}$$

where R_n denotes the local window, z_m symbolizes the gray level of R_n, R_s indicates cardinality of R_n and $\overline{z_n}$ denotes the mean value of gray level. The filtered image is incorporated in Equation 9.7.

The exponential of LIV gives the weight function.

$$\alpha_n = \exp\left(\sum_{m \in R_n,\ n \neq m} \mathrm{LIV}_n\right) \tag{9.8}$$

$$w_n = \frac{\alpha_n}{\sum_{m \in R_n} \alpha_m} \tag{9.9}$$

Express the regularization kernel factor as follows

$$\beta_n = \begin{cases} 2 + w_n, & \overline{z_n} < z_n \\ 2 - w_n, & \overline{z_n} > z_n \\ 0, & \overline{z_n} = z_n \end{cases} \tag{9.10}$$

From the above equation, it is concluded that β_n is directly proportional to LIV. For the minimization of computational time, the value of β_n has to be computed before the clustering process.

Replace the Euclidean distance term $\left\| z_n - V_j^2 \right\|$ by $\left\| \beta(z_n) - \beta(V_j)^2 \right\|$ which is illustrated as follows.

$$\left\| \beta(z_n) - \beta(V_j) \right\|^2 = A(z_n, z_n) + A(V_j, V_j) - 2A(z_n, V_j) \tag{9.11}$$

where A denotes the kernel function.

The gaussian radial basis function (GRBF) kernel is

$$A(z_n, V_j) = \exp\left(\frac{-z_n - V_j^2}{2\sigma^2}\right) \tag{9.12}$$

where σ indicates the width of the kernel.

$$\sigma = \left[\frac{\sum_{n=1}^{R}(d_n - \bar{d})^2}{R-1}\right]^{1/2} \tag{9.13}$$

where $d_n = z_n - \bar{z}$ is the distance from gray level of image element n to the gray level mean of all image elements and $\bar{d}$ is the average of all distances. Using GRBF, the kernel function is described as follows.

$$\beta(z_n) - \beta(V_j)^2 = 2\left[1 - A(z_n, V_j)\right] \tag{9.14}$$

The objective function of the proposed ARKFCM is as follows.

$$J_{\text{ARKFCM}} = 2\left\{\sum_{n=1}^{R}\sum_{j=1}^{c} U_{nj}^{f}\left[1 - A(z_n, V_j)\right] + \sum_{n=1}^{R}\sum_{j=1}^{c} \beta_n U_{nj}^{f}\left[1 - A(\bar{z}_n, V_j)\right]\right\} \tag{9.15}$$

$$U_{nj} = \frac{\left\{\left[1 - A(z_n, V_j)\right] + \beta_n\left[1 - A(\bar{z}_n, V_j)\right]\right\}^{\frac{-1}{(f-1)}}}{\sum_{m=1}^{c}\left\{\left[1 - A(z_n, V_m)\right] + \beta_n\left[1 - A(\bar{z}_n, V_m)\right]\right\}^{\frac{-1}{(f-1)}}} \tag{9.16}$$

$$V_j = \frac{\sum_{n=1}^{R} U_{nj}^{f}\left[A(z_n, V_j) z_n + \beta_n A(\bar{z}_n, V_j)\bar{z}_n\right]}{\sum_{n=1}^{R} U_{nj}^{f}\left[A(z_n, V_j) + \beta_n A(\bar{z}_n, V_j)\right]} \tag{9.17}$$

The CSO was incorporated in the ARKFCM algorithm for the cluster centroids selection rather than the random initialization. The different steps in the ROI extraction concerning the ARKFCM algorithm are represented in Equations 9.7–9.17. The different stages in the CSO are represented in Equations 9.1–9.6. The CSO was incorporated in the cluster centroid initialization represented in Equation 9.17. The parameters of the clustering and optimization technique are initialized and the first stage is the preprocessing of the input image. The LIV is framed from the filtered image and is used in the formulation of the Gaussian kernel. The Gaussian kernel is incorporated in the clustering algorithm, where the CSO is used in the cluster centroids initialization.

9.3 RESULTS AND DISCUSSION

The algorithms are developed in MATLAB® 2010a and tested on real-time corona virus disease (COVID-19) CT images. The source of data is as follows [23,24]: prior permission was obtained from the Radiological Society of North America for the usage of real-time COVID 19 CT images in the research work. This chapter proposes two segmentation algorithms for the extraction of ROI: ARKFCM and ARKFCM-crow techniques. The medical images are prone to noise, and here two filtering techniques are used before segmentation. The parameters of the algorithms are as follows: the cluster number is changed from $C=3$ to $C=5$ and the size of the local window chosen is 3. The parameters of the crow optimization are awareness probability (Ap = 0.1), flight length ($f_l=0.1$), population value (np = 10), and number of iterations (iter = 10). The values are fixed based on the analysis of previously obtained results on medical images [25]. The CSO was utilized in the optimum cluster centroids selection in fuzzy clustering for the ROI extraction on abdomen CT images [25].

The CSO was used in the adaptive regularized kernel FCM for the cluster centroids initialization. The input images are depicted in Figure 9.2.

The ARKFCM segmentation results corresponding to the median filter for various cluster values of $C=3$, 4, and 5 are depicted in Figures 9.3–9.5. The ARKFCM-crow segmentation results corresponding to the median filter for various cluster values of $C=3$, 4, and 5 are depicted in Figures 9.6–9.8. The performance metrics validation reveals that the median filter when coupled with the segmentation algorithm was found to be superior when equated with the average filter. For performance

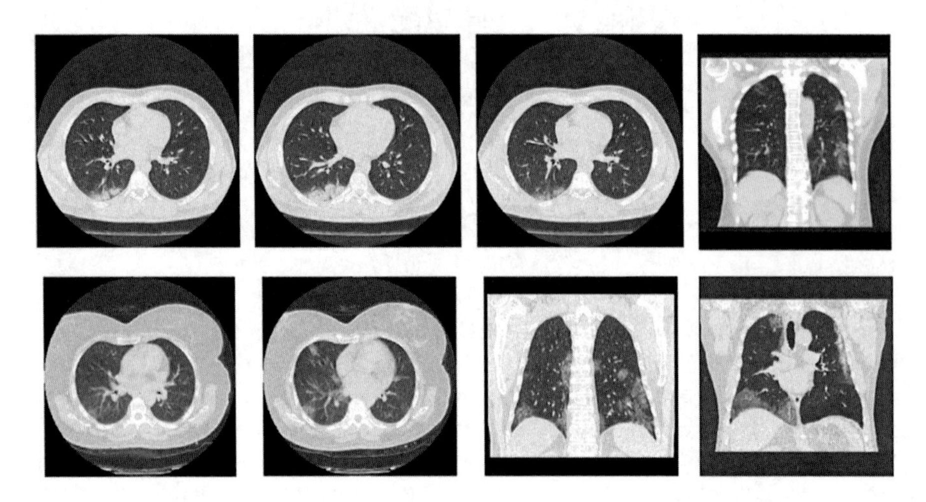

FIGURE 9.2 Input COVID 19 CT images (ID1–ID8).

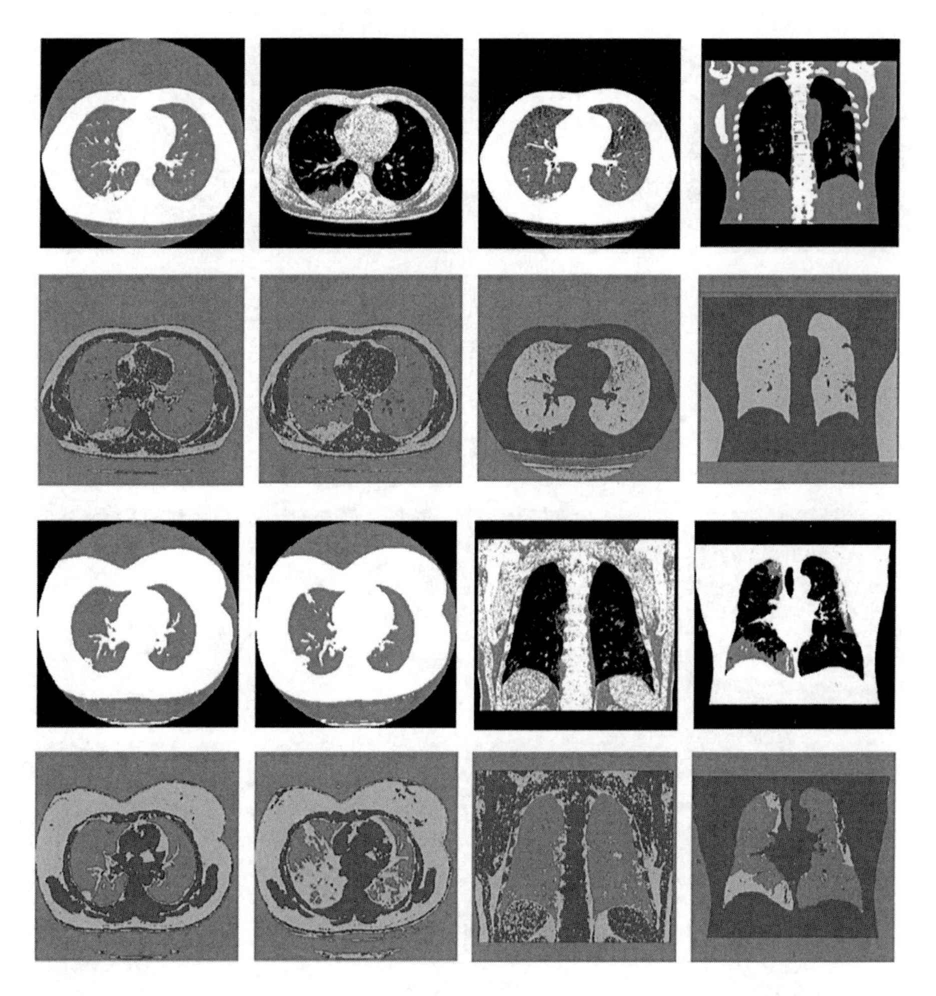

FIGURE 9.3 Adaptively regularized kernel-based fuzzy c-means (ARKFCM) segmentation results for the cluster value of 3 corresponding to the input images in Figure 9.2 (ID1–ID8).

validation, the following cluster validity metrics [26] are used. In Figures 9.3–9.8, the first and third row depicts the gray-scale segmentation result, the second and the fourth row depicts the colored segmentation results. Corresponding to the cluster number, the ROI is depicted in different colors.

The expression for partition coefficient and partition entropy is as follows: maximum value of partition coefficient (PC) and minimum value of partition entropy (PE) favors the efficiency of the clustering algorithm.

$$V_{\mathrm{PC}}(C) = \frac{1}{N}\sum_{c=1}^{C}\sum_{i=1}^{N} F_{ic}^{2} \tag{9.18}$$

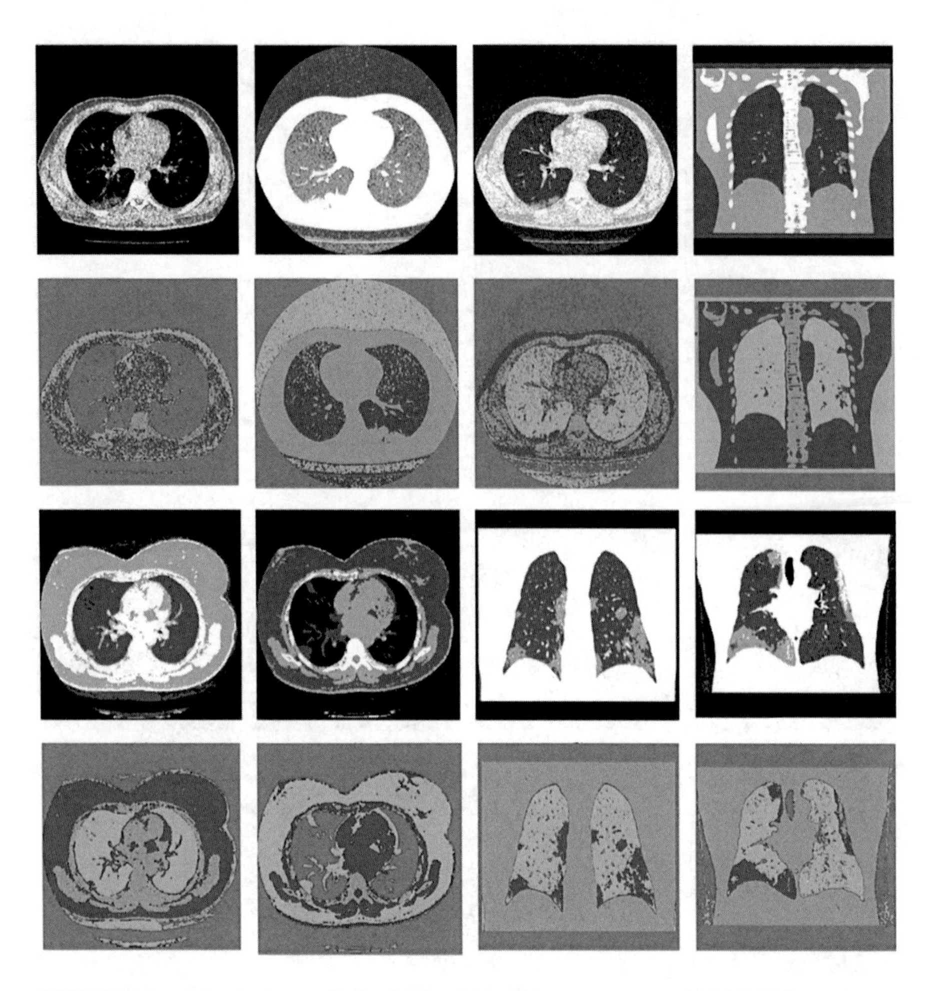

FIGURE 9.4 Adaptively regularized kernel-based fuzzy c-means (ARKFCM) segmentation results for cluster value of 4 corresponding to the input images in Figure 9.2 (ID1–ID8).

$$V_{\text{PE}}(C) = \frac{1}{N}\sum_{c=1}^{C}\sum_{i=1}^{N} F_{ic}\log_2(F_{ic}) \tag{9.19}$$

The expression for Xie and Beni Index (XBI) and Fukuyama and Sugeno Index (FSI) are as follows

$$V_{\text{XBI}}(C) = \frac{\sum_{c=1}^{C}\sum_{i=1}^{N} F_{ic}^{C}\, y_i - v_c^{\,2}}{N.\min\limits_{i \neq j} y_i - v_j^{\,2}} \tag{9.20}$$

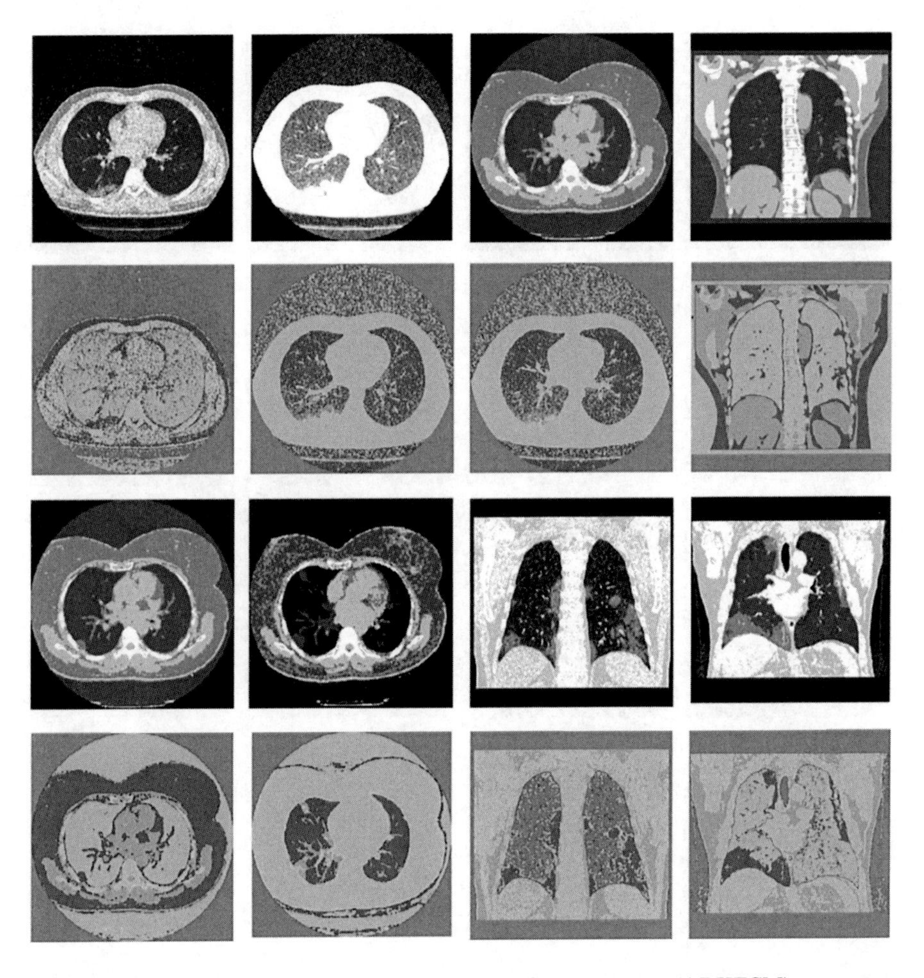

FIGURE 9.5 Adaptively regularized kernel-based fuzzy c-means (ARKFCM) segmentation results for cluster value of 5 corresponding to the input images in Figure 9.2 (ID1–ID8).

$$V_{\text{FSI}}(C) = \sum_{c=1}^{C}\sum_{i=1}^{N} F_{ic}^{m}\, y_i - v_c{}^2 - \sum_{c=1}^{C}\sum_{i=1}^{N} F_{ic}^{m}\, v_c - \hat{v}^2 \tag{9.21}$$

The minimum value of XBI and FSI indicates the efficiency of the clustering algorithm. The performance metrics were determined for various cluster values and the results are represented in Tables 9.1–9.5.

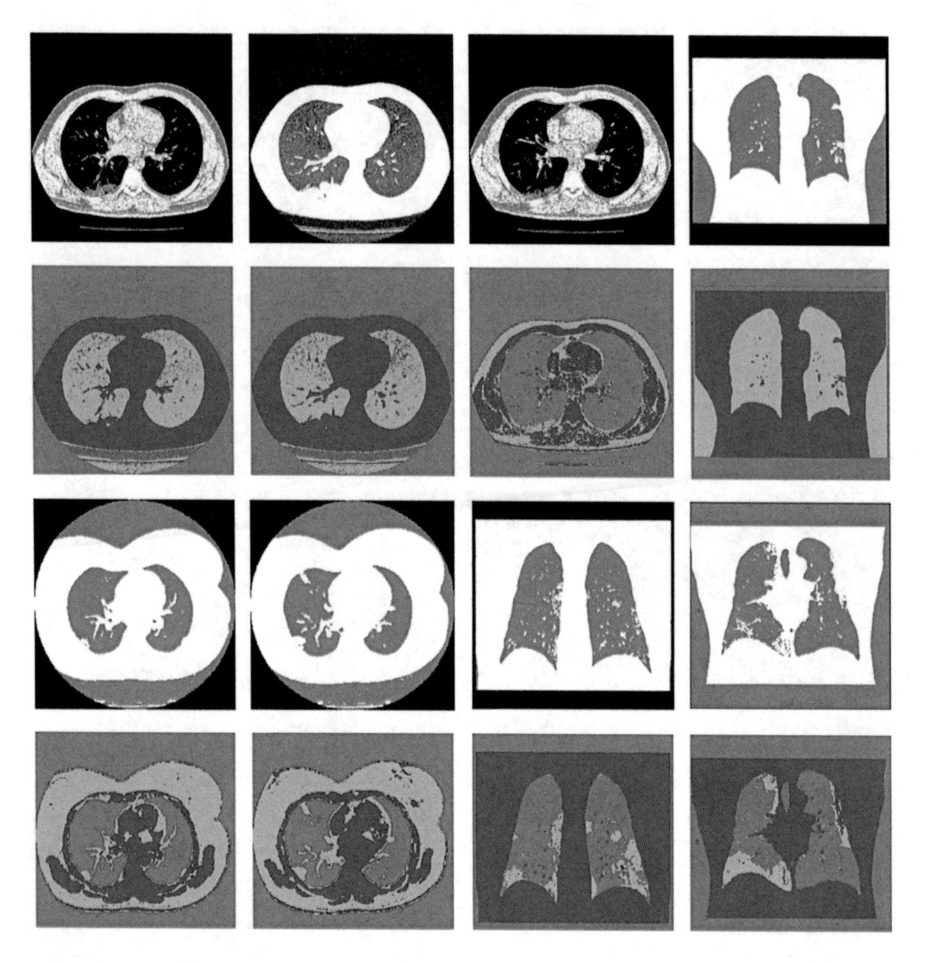

FIGURE 9.6 Adaptively regularized kernel-based fuzzy c-means (ARKFCM)-crow segmentation results for cluster value of 3 corresponding to the input images in Figure 9.2 (ID1–ID8).

For the FCM clustering algorithm, the minimum cluster value is 2 and it can be assigned the values of 2, 3, 4, and 5. In this research work, cluster values were taken as 3, 4, and 5 and validated by the performance metrics. The cluster value can be decided by the values of performance metrics. From the performance metrics values, it is evident that a cluster value of 3 is appropriate (Table 9.6).

The PC and PE values are the primary metrics for deciding the cluster value. For an efficient clustering algorithm, the PC value should be high and the PE value needs to be low. The results in Table 9.7 reveal that for the cluster value of 3 the PC is high and PE is low. Apart from PC and PE, the clustering result was also validated by the other performance metrics (Figure 9.9).

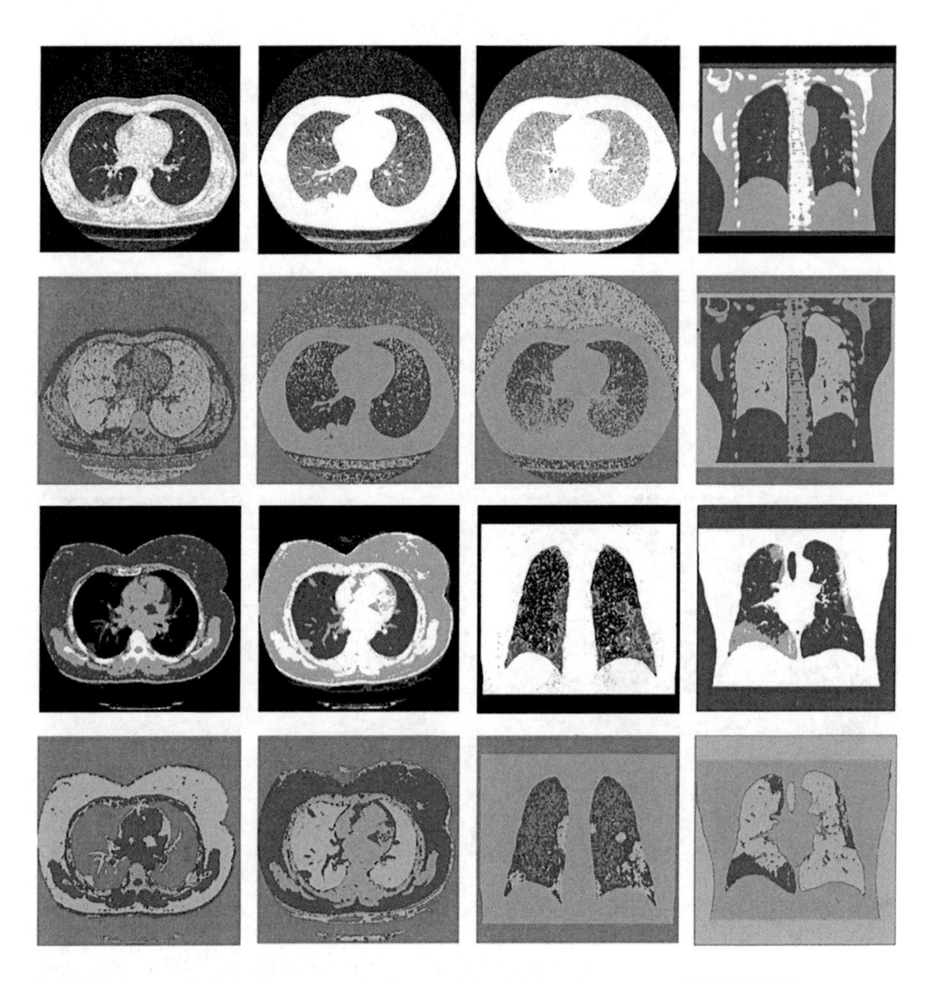

FIGURE 9.7 Adaptively regularized kernel-based fuzzy c-means (ARKFCM)-crow segmentation results for cluster value of 4 corresponding to the input images in Figure 9.2 (ID1–ID8).

The performance metrics ideal values for a clustering algorithm is difficult to highlight exactly; however, for an efficient clustering algorithm, the PC value should be high preferably greater than 0.8 and the PE value should be low. The XBI and FSI values also should be low for an efficient clustering algorithm. The metrics are represented as PC^{-}, PE^{-}, XBI^{-}, and FSI^{-} indicating their nature. The statistical analysis was performed and the critical difference diagrams are plotted for the metrics. Since the ARKFCM-crow optimization segmentation algorithm was found to be proficient, the segmentation result for eight more datasets is depicted in Figure 9.10.

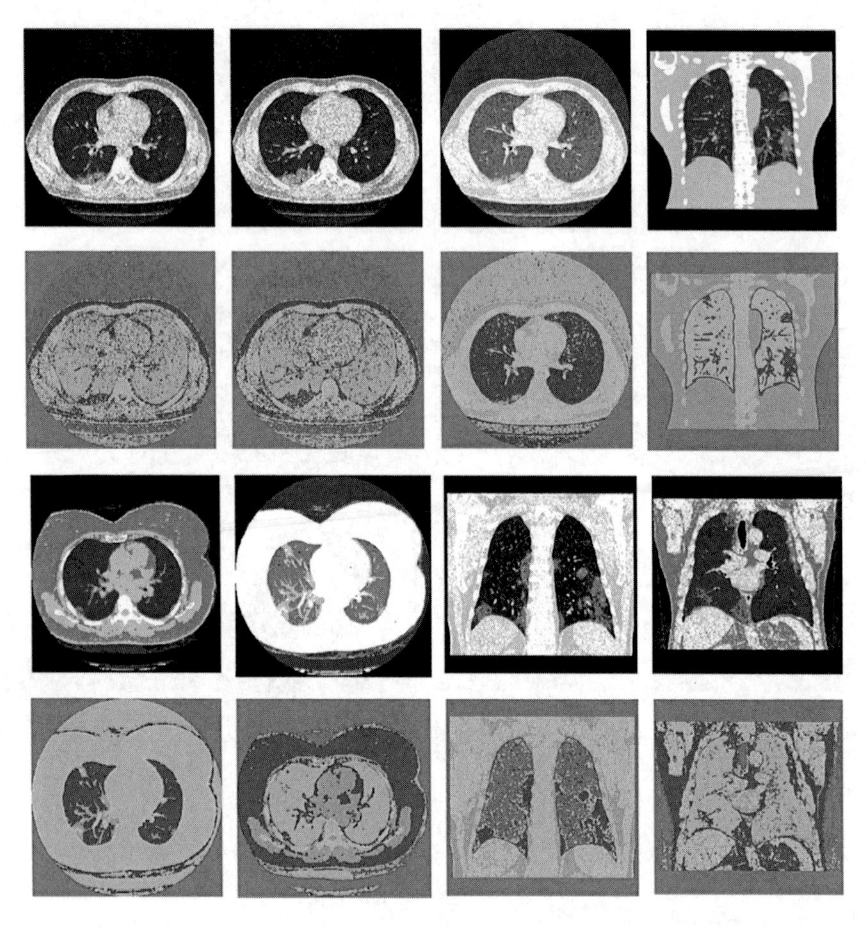

FIGURE 9.8 Adaptively regularized kernel-based fuzzy c-means (ARKFCM)-crow segmentation results for cluster value of 5 corresponding to the input images in Figure 9.2 (ID1–ID8).

TABLE 9.1
Segmentation Metrics for Cluster Value of 3 with the Median Filter

Image Details	ARKFCM-Crow				ARKFCM			
	PC⁺	PE⁻	XBI⁻	FSI⁻	PC⁺	PE⁻	XBI⁻	FSI⁻
ID 1	0.82	0.32	−4.01	−1.41e+09	0.80	0.33	−0.04	−1.05e+09
ID 2	0.82	0.32	−3.65	−1.41e+09	0.80	0.33	−0.04	−1.06e+09
ID 3	0.80	0.31	−0.04	−1.05e+09	0.82	0.32	−4.11	−1.43e+09
ID 4	0.87	0.20	−0.01	−2.29e+10	0.87	0.21	−0.01	−2.29e+10
ID 5	0.80	0.34	−2.38	−1.84e+08	0.80	0.35	−2.38	−1.84e+08
ID 6	0.88	0.35	−1.84	−4.61e+08	0.80	0.37	−2.20	−2.02e+08
ID 7	0.86	0.30	−2.90	−6.26e+08	0.82	0.31	−0.15	−8.13e+08
ID 8	0.91	0.16	−0.03	−1.25e+09	0.91	0.17	−0.03	−1.25e+09

TABLE 9.2
Segmentation Metrics for Cluster Value of 4 with the Median Filter

Image Details	ARKFCM-Crow				ARKFCM			
	PC+	PE-	XBI-	FSI-	PC+	PE-	XBI-	FSI-
ID 1	0.76	0.45	−7.80	−1.72e+09	0.72	0.48	−1.65	−1.04e+09
ID 2	0.88	0.22	−0.02	−2.00e+09	0.71	0.53	−0.03	−1.20e+09
ID 3	0.72	0.48	−1.69	−1.04e+09	0.63	0.67	−5.05	3.86e+08
ID 4	0.82	0.31	−0.15	−2.02e+10	0.82	0.31	−0.15	−2.02e+10
ID 5	0.78	0.40	−1.18	−2.02e+08	0.77	0.45	−1.31	−2.44e+08
ID 6	0.76	0.45	−1.25	−2.57e+08	0.77	0.41	−1.18	−2.03e+08
ID 7	0.92	0.16	−0.02	−2.61e+09	0.75	0.47	−0.12	−1.05e+09
ID 8	0.88	0.23	−0.02	−1.33e+09	0.90	0.21	−0.02	−1.62e+09

TABLE 9.3
Segmentation Metrics for Cluster Value of 5 with the Median Filter

Image Details	ARKFCM-Crow				ARKFCM			
	PC+	PE-	XBI-	FSI-	PC+	PE-	XBI-	FSI-
ID 1	0.68	0.58	−5.13	−1.16e+09	0.68	0.58	−5.13	−1.16e+09
ID 2	0.67	0.64	−0.03	−1.26e+09	0.68	0.58	−4.90	−1.15e+09
ID 3	0.67	0.64	−0.03	−1.24e+09	0.83	0.30	−0.78	−1.79e+09
ID 4	0.81	0.34	−0.15	−2.07e+10	0.78	0.43	−0.17	−1.55e+10
ID 5	0.84	0.31	−0.22	−4.69e+08	0.76	0.46	−0.68	−2.24e+08
ID 6	0.69	0.61	−1.20	−2.58e+08	0.90	0.21	−0.02	−4.65e+08
ID 7	0.68	0.59	−1.40	−7.70e+08	0.68	0.59	−1.40	−7.70e+08
ID 8	0.85	0.28	−0.18	−1.23e+09	0.79	0.37	−0.82	−1.32e+09

TABLE 9.4
Segmentation Metrics for Cluster Value of 3 with the Average Filter

Image Details	ARKFCM-Crow				ARKFCM			
	PC+	PE-	XBI-	FSI-	PC+	PE-	XBI-	FSI-
ID 1	0.82	0.32	−4.01	−1.41e+09	0.80	0.33	−0.04	−1.06e+09
ID 2	0.82	0.32	−4.01	−1.41e+09	0.82	0.32	−3.65	−1.41e+09
ID 3	0.80	0.33	−0.04	−1.05e+09	0.93	0.13	−0.02	−1.87e+09
ID 4	0.64	0.60	−3.32	1.16e+10	0.83	0.31	−0.62	−1.47e+10
ID 5	0.58	0.70	−8.24	3.41e+08	0.80	0.35	−2.38	−1.84e+08
ID 6	0.80	0.35	−2.20	−2.02e+08	0.84	0.29	−0.06	−2.37e+08
ID 7	0.95	0.10	−0.01	−2.77e+09	0.82	0.31	−0.15	−8.13e+08
ID 8	0.89	0.19	−0.01	−1.42e+09	0.89	0.19	−0.47	−1.48e+09

TABLE 9.5
Segmentation Metrics for Cluster Value of 4 with the Average Filter

Image Details	ARKFCM-Crow				ARKFCM			
	PC^+	PE^-	XBI^-	FSI^-	PC^+	PE^-	XBI^-	FSI^-
ID 1	0.73	0.47	−2.67	−1.00e+09	0.73	0.49	−0.04	−1.08e+09
ID 2	0.73	0.49	−0.04	−1.09e+09	0.73	0.47	−2.42	−9.98e+08
ID 3	0.86	0.25	−1.43	−1.83e+09	0.76	0.45	−16.10	−1.68e+09
ID 4	0.80	0.39	−0.75	−1.49e+10	0.90	0.20	−0.01	−2.55e+10
ID 5	0.85	0.27	−0.43	−4.09e+08	0.92	0.16	−0.01	−4.82e+08
ID 6	0.85	0.28	−0.45	−4.25e+08	0.78	0.41	−0.14	−2.14e+08
ID 7	0.72	0.49	−2.32	−6.17e+08	0.71	0.54	−6.01	−6.44e+08
ID 8	0.84	0.27	−0.27	−1.26e+09	0.89	0.22	−0.01	−1.93e+09

TABLE 9.6
Segmentation Metrics for Cluster Value of 5 with the Average Filter

Image Details	ARKFCM-Crow				ARKFCM			
	PC^+	PE^-	XBI^-	FSI^-	PC^+	PE^-	XBI^-	FSI^-
ID 1	0.86	0.26	−0.02	−1.89e+09	0.83	0.29	−1.21	−1.71e+09
ID 2	0.69	0.57	−9.40	−1.12e+09	0.86	0.26	−0.02	−1.91e+09
ID 3	0.67	0.61	−2.38	−8.62e+08	0.86	0.26	−0.02	−1.88e+09
ID 4	0.88	0.25	−0.01	−2.35e+10	0.81	0.34	−0.15	−2.07e+10
ID 5	0.76	0.47	−0.84	−2.21e+08	0.90	0.21	−0.02	−4.55e+08
ID 6	0.74	0.49	−1.08	−1.94e+08	0.88	0.23	−0.25	−4.30e+08
ID 7	0.82	0.34	−0.81	−2.16e+09	0.84	0.28	−0.35	−2.14e+09
ID 8	0.85	0.27	−0.18	−1.24e+09	0.88	0.24	−0.02	−1.66e+09

TABLE 9.7
Segmentation Metrics of ARKFCM-Crow Algorithm for Cluster Value of 3 with the Median Filter

	ARKFCM-Crow			
Image Details	PC	PE	XBI	FSI
ID9	0.8422	0.2963	−1.3117	−1.477154e+09
ID10	0.7851	0.3720	−0.0589	−8.567117e+08
ID11	0.7862	0.3692	−0.0488	−9.602068e+08
ID12	0.8220	0.3267	−1.5583	−1.320702e+09
ID13	0.9294	0.1425	−0.0484	−1.897507e+09
ID14	0.8724	0.2373	−0.4662	−3.400079e+09
ID15	0.9482	0.1067	−0.0105	−3.842347e+09
ID16	0.8810	0.2124	−0.0057	−2.817937e+10
ID17	0.9253	0.1442	−0.3382	−3.349547e+10

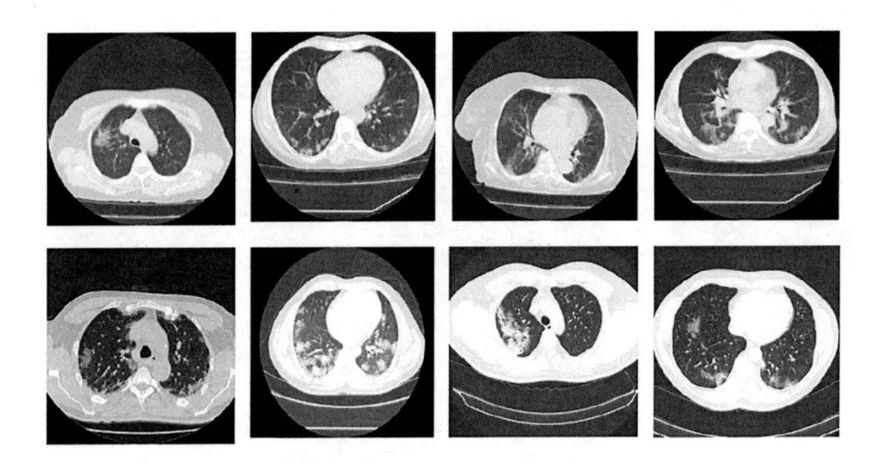

FIGURE 9.9 Input COVID-19 CT images (ID9–ID17).

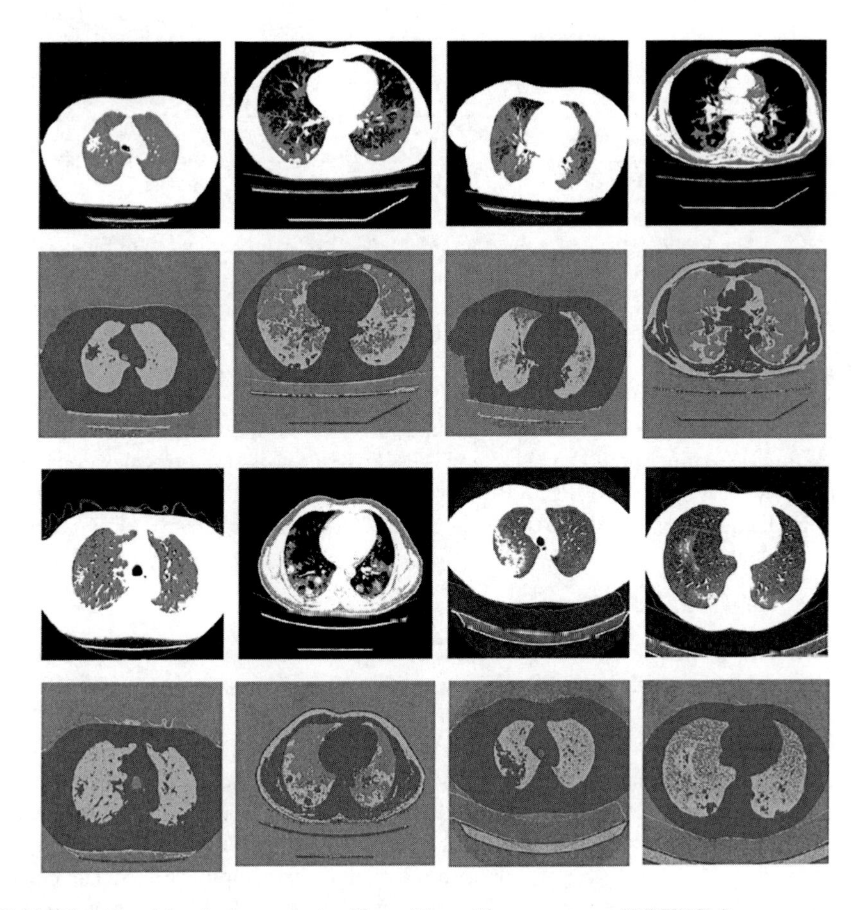

FIGURE 9.10 Adaptively regularized kernel-based fuzzy c-means (ARKFCM)-crow segmentation results for cluster value of 5 corresponding to the input images in Figure 9.2 (ID9–ID17).

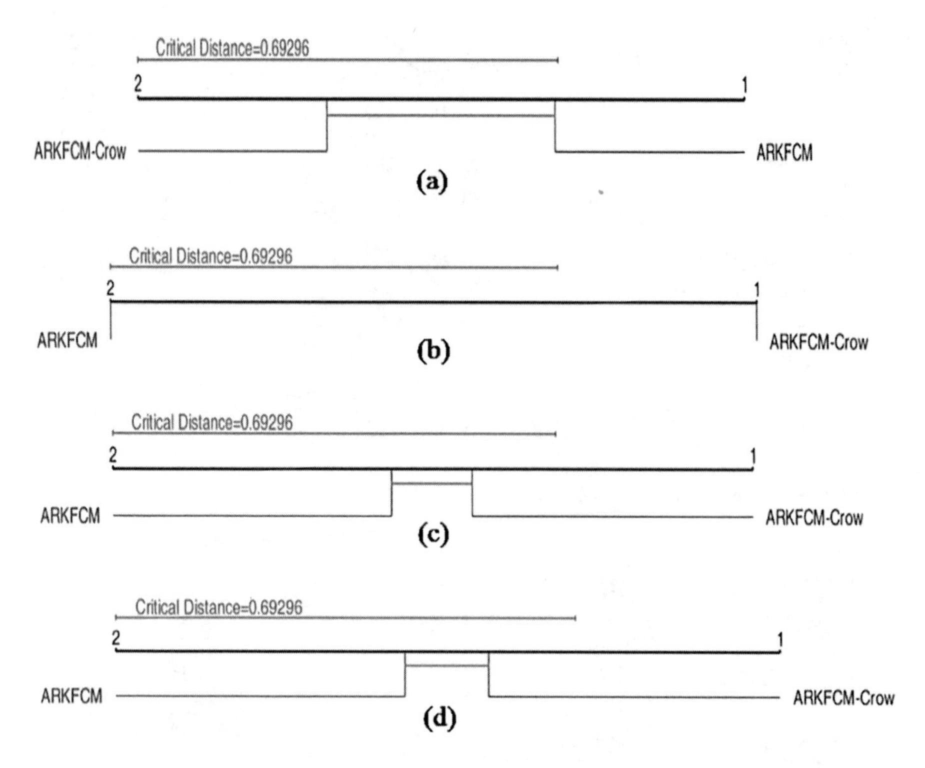

FIGURE 9.11 Critical difference diagram for (a) PC, (b) PE, (c) Xie and Beni Index (XBI), and (d) Fukuyama and Sugeno Index metrics.

The parallel processing improves the computational efficiency and FCM was implemented in GPU [27]. The critical difference diagram [28] for the clustering algorithms is depicted in Figure 9.11 for the performance metrics. The plots reveal that in terms of PC, ARKFCM-crow is superior when compared with the ARKFCM algorithm (PC value should be high), and ARKFCM-crow is superior when compared with ARKFCM in terms of PE, XBI, and FSI (values should be low). The critical difference diagram is a visual representation of the statistical significance of parameters.

9.4 CONCLUSION

This chapter proposes ARKFCM and ARKFCM-crow segmentation algorithms for the ROI extraction in COVID-19 CT images. The ARKFCM was found to be efficient when compared with the classical clustering technique and the crow optimization was further coupled with the ARKFCM for the selection of cluster centroids. Before segmentation, preprocessing was done by the median filter and the performance validation by cluster validity metrics reveals that ARKFCM-crow with preprocessing by the median filter was proficient in the ROI extraction. The less parameter tuning makes the CSO an attractive algorithm for engineering applications. The cluster values are manually initialized in this work and based on the performance metrics values, the cluster value is finalized. The future work will be an emphasis on the automatic selection of cluster values and hardware implementation of the proposed clustering technique in an embedded processor for telemedicine application.

REFERENCES

1. Narkhede H.P. Review of image segmentation techniques. *International Journal of Science and Modern Engineering*. 2013;1(8):54–61.
2. Kumar S.N., Fred A.L., and Varghese P.S. An overview of segmentation algorithms for the analysis of anomalies on medical images. *Journal of Intelligent Systems*. 2018;29(1):612–625.
3. Gautam A., Sadhya D., Raman B. A Modified FCM-Based Brain Lesion Segmentation Scheme for Medical Images. In *Proceedings of 3rd International Conference on Computer Vision and Image Processing* 2020 (pp. 149–159). Springer, Singapore.
4. Cai J. Segmentation and diagnosis of liver carcinoma based on adaptive scale-kernel fuzzy clustering model for CT images. *Journal of Medical Systems*. 2019;43(11):322.
5. Elazab A., Wang C., Jia F., Wu J., Li G., and Hu Q. Segmentation of brain tissues from magnetic resonance images using adaptively regularized kernel-based fuzzy-means clustering. *Computational and Mathematical Methods in Medicine*. 2015;2015.
6. Debakla M., Salem M., Djemal K., and Benmeriem K. Fuzzy farthest point first method for MRI brain image clustering. *IET Image Processing*. 2019;13(13):2395–2400.
7. Fang R., Lu Y., Liu X., Liu Z. Segmentation of Brain MR Images Using an Adaptively Regularized Kernel FCM Algorithm with Spatial Constraints. In *2017 10th International Congress on Image and Signal Processing, BioMedical Engineering and Informatics (CISP-BMEI)* 2017 (pp. 1–6). IEEE, New York.
8. Hu G., and Du Z. Adaptive kernel-based fuzzy c-means clustering with spatial constraints for image segmentation. *International Journal of Pattern Recognition and Artificial Intelligence*. 2019;33(01):1954003.
9. Song J., and Zhang Z. A modified robust FCM model with spatial constraints for brain MR image segmentation. *Information*. 2019;10(2):74.
10. Nadernejad E., Sharifzadeh S. A new method for image segmentation based on Fuzzy C-means algorithm on pixonal images formed by bilateral filtering. *Signal, Image and Video Processing*. 2013;7(5):855–863.
11. Elazab A., AbdulAzeem Y.M., Wu S., and Hu Q. Robust kernelized local information fuzzy C-means clustering for brain magnetic resonance image segmentation. *Journal of X-Ray Science and Technology*. 2016;24(3):489–507.
12. Gomathi M., and Thangaraj P. A new approach to lung image segmentation using fuzzy possibilistic C-means algorithm. *arXiv* preprint arXiv:1004.1768. 2010.
13. Taher F., and Sammouda R. Lung cancer detection by using artificial neural network and fuzzy clustering methods. In *2011 IEEE GCC Conference and Exhibition (GCC)* 2011 (pp. 295–298). IEEE, New York.
14. Harish B.S., Kumar S.A., Masulli F., and Rovetta S. Adaptive initialization of cluster centers using ant colony optimization: Application to medical images. In *ICPRAM* 2017 (pp. 591–598). SciTePress, Portugal.
15. Liu H., Zhang C.M., Su Z.Y., Wang K., Deng K. Research on a pulmonary nodule segmentation method combining fast self-adaptive FCM and classification. *Computational and Mathematical Methods in Medicine*. 2015;2015.
16. Raza K. Fuzzy logic based approaches for gene regulatory network inference. *Artificial Intelligence in Medicine*. 2019;97:189–203.
17. Zhang Y., Wu S., Yu G., and Wang D. A hybrid image segmentation approach using watershed transform and FCM. In *Fourth International Conference on Fuzzy Systems and Knowledge Discovery (FSKD 2007)* 2007 (Vol. 4, pp. 2–6). IEEE, New York.

18. Raza K., and Singh N.K. A tour of unsupervised deep learning for medical image analysis. *arXiv* preprint arXiv:1812.07715. 2018.
19. Kavitha P., and Prabakaran S. A novel hybrid segmentation method with particle swarm optimization and fuzzy C-mean based on partitioning the image for detecting lung cancer. Preprints 2019, 2019060195.
20. Askarzadeh A. A novel metaheuristic method for solving constrained engineering optimization problems: Crow search algorithm. *Computers & Structures*. 2016;169:1–2.
21. Gopal N.N., and Karnan M. Diagnose brain tumor through MRI using image processing clustering algorithms such as Fuzzy C Means along with intelligent optimization techniques. In *2010 IEEE International Conference on Computational Intelligence and Computing Research* 2010 (pp. 1–4). IEEE, New York.
22. Omran M.G., Salman A., and Engelbrecht A.P. Dynamic clustering using particle swarm optimization with application in image segmentation. *Pattern Analysis and Applications*. 2006;8(4):332.
23. Shi H., Han X., and Zheng C. Evolution of CT manifestations in a patient recovered from 2019 novel coronavirus (2019-nCoV) pneumonia in Wuhan, China. *Radiology*. 2020;295(1): 20.
24. Duan Y.N., and Qin J. Pre-and posttreatment chest CT findings: 2019 novel coronavirus (2019-nCoV) pneumonia. *Radiology*. 2020;295(1): 21.
25. Fred A.L., Kumar S.N., Padmanaban P., Gulyas B., and Kumar H.A. Fuzzy-Crow search optimization for medical image segmentation. In *Applications of Hybrid Metaheuristic Algorithms for Image Processing* 2020 (pp. 413–439). Springer, Cham.
26. Pakhira M.K., Bandyopadhyay S., and Maulik U. A study of some fuzzy cluster validity indices, genetic clustering and application to pixel classification. *Fuzzy Sets and Systems*. 2005;155(2):191–214.
27. Al-Ayyoub M., Abu-Dalo A.M., Jararweh Y., Jarrah M., Al Sa'd M. A gpu-based implementations of the fuzzy c-means algorithms for medical image segmentation. *The Journal of Supercomputing*. 2015;71(8):3149–3162.
28. Demšar J. Statistical comparisons of classifiers over multiple data sets. *Journal of Machine Learning Research*. 2006;7(Jan):1–30.

10 Estimating the Effect of Social Distancing in the Progression Dynamics of COVID-19

Narender Kumar
University of Delhi
Jamia Millia Islamia

Shweta Sankhwar
Babasaheb Bhimrao Ambedkar University

Ravins Dohare
Jamia Millia Islamia

CONTENTS

10.1 INTRODUCTION

There are numerous events in history which exhibit that various diseases had serious impact on humans. Of these examples, the most well-known is the Black Death in Europe in the fourteenth century [1,2]. The prevalence of such infectious diseases still exists in the human population. More specifically, viral infectious diseases hold major share among other infectious diseases [3]. The outbreak of such notable viral infectious diseases have been observed and marked time-to-time in the world. In recent times, several emerging viral diseases such as Zika, Ebola, chikungunya, dengue, severe acute respiratory syndrome (SARS), Middle East respiratory syndrome coronavirus (MERS), and SARS-CoV-2 (severe acute respiratory syndrome coronavirus 2), are a serious threat to humans in terms of health and wealth. To eradicate or cease the outbreak of such infectious diseases in the future is a major challenge for the entire scientific communities. If we understand the phenomena of disease propagation in a given human population, then we will be able to equip ourselves for preventing it through immunization, quarantine, or by other control strategies. The transmission characteristics are not identical for all viral infectious diseases. Some viral diseases spread by close contact with infected individuals to other susceptible individuals in the population, for example, influenza, measles, and chickenpox. On the other hand, some viral infectious diseases transmit from one individual to another through a biological vector, such as Zika, chikungunya, dengue, and West Nile. Certain diseases are highly contagious in comparison to other diseases, for example, measles and influenza are more contagious than Zika, dengue, or chikungunya. Some diseases like mumps and measles confer a lifelong immunity, whereas influenza has a short-term immunity. Therefore, based on the transmission characteristics, we need different control strategies for different viral diseases. In the series of infectious diseases outbreak, a recent outbreak of life-threatening disease COVID-19 caused by the SARS-CoV-2 posed a health emergency worldwide [4].

Many researchers tried to capture the transmission characteristic of the COVID-19 using the various types of mathematical model based on different approaches. Ivorra et al. [5] formulated a mathematical model of COVID-19 based on different sanitary conditions and infectiousness of the hospitalized individuals. A good agreement between actual and estimated cases is shown by their model. They estimated the behavior of the model output using truncation technique when the reported data were incomplete. Through the model, they also calculated the number of beds required in the hospitals. Cao et al. [6] developed mathematical model with the consideration of prevention and control measures for the transmission of COVID-19. They used six compartments and time series based on the technique in their model. On the basis of time series and kinetic model analysis, they explained that number of commutative cases of COVID-19 in mainland China can reach 36,343 in a week. Kucharski et al. [7] used stochastic modeling to estimate the transmission dynamics of COVID-19 in Wuhan. They also estimated how transmission had varied internationally during the time period from January 2020 to February 2020. Shaikh et al. [8] proposed a Bats-Hosts-Reservoir-People transmission model based on fractional order derivatives for COVID-19. In this work, they simulated the potential transmission with individual, social response, and control measures opted by the government. They estimated various model parameters which were based on the actual outbreak data from March 14 to March 26, 2020. They also did the stability analysis of the mathematical model through Picard successive approximation technique and Banach's fixed point theory. Khan et al. [9] also developed a model based on fractional order derivatives. They assumed the infection in human population thorough seafood market as an infection reservoir. The seafood market was considered as infection reservoir when the virus came here through bats or unknown host (may be wild animals). They considered infected humans as symptomatically or asymptomatically in their model. The model was fitted on the actual data outbreak from January 21 to January 28, 2020, and estimated the parameters involved in the model. All these models can be of interest for policy makers to prevent the transmission of COVID-19 in many countries or states.

Thus, mathematical models are used as a crucial tool to investigate the transmission dynamics of infectious diseases. The value of some parameters involved in the model can be identified based on clinical research, and some of these may be estimated using some computational techniques. These computational techniques cover simple/tedious mathematical calculations or complex computer simulations based on different approaches such as statistical techniques, machine learning, data mining, or network simulation. Based on these parameter estimations, the progression dynamics of the disease can be investigated in a futuristic scenario. Outcomes from these mathematical models may be useful to design the control strategies and their impact; apart from this, it also enables insight into the vaccination condition in the population. The vaccine for COVID-19 has not been yet discovered. Therefore, the quarantine is a control measure in the major part of the affected world for the recent outbreak of COVID-19 [10]. Many countries announced the lock-down for many days to fight against COVID-19 [11]. The social distancing measure played a crucial role in the mitigation of COVID-19 in several geographical regions. In this chapter, we have formulated a mathematical model through a proposed susceptible, exposed, asymptomatic infected, symptomatic infected, hospitalized, recovered, and Death (SEAIJRD) compartmental modeling in which the whole is population assumed to be subcategorized on the basis of infection status of individuals. This formulated model has been applied on the current outbreak data of COVID-19 in India. Using the COVID-19 real outbreak data, some unknown parameters are estimated for this outbreak. On the basis of these estimated parameters, a total number of infected individuals are predicted. Furthermore, we studied the effect of social distancing on the transmission of COVID-19. For this purpose, we assumed the social distancing in term of average contact rate (c). The effect of this average contact rate on the COVID-19 transmission has been estimated. The basic reproduction number (R_0) is also computed for different average contact rates. The sensitivity for the basic reproduction number is also discussed with respect to parameters involved in its expression. This study may be helpful to design or implement control strategy for COVID-19 in India or any other geographical region in the current scenario or future perspective.

10.2 MATERIAL AND METHODS

10.2.1 Mathematical Model

We formulated a mathematical model for COVID-19 as shown in Figure 10.1. In the proposed model, the whole population is subdivided into seven mutually exclusive classes based on the infectious status of a person in the given population. These mutually exclusive classes are denoted by the letters S, E, A, I, J, R, D and represent susceptible (healthy but on risk for disease), exposed (infected person but not infectious), asymptomatic (infected person without symptoms), infected, hospitalized (mild, moderate and severe cases), recovered, and dead categories, respectively. The number of individuals move from one compartment to another with respect to time. The direction

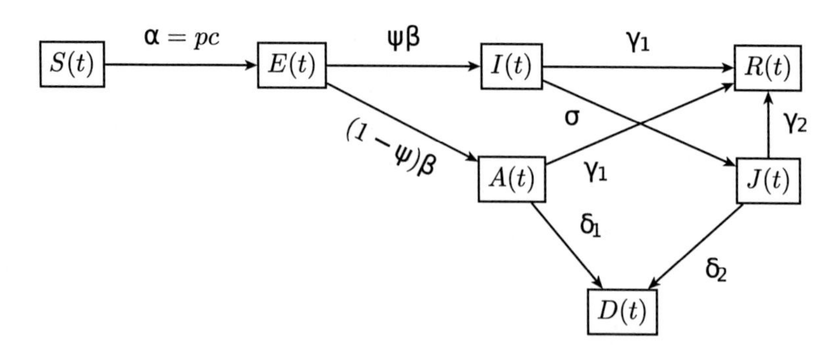

FIGURE 10.1 The schematic diagram of the SEAIJRD-type transmission model. *S, E, I, A, J, R* and *D* denote susceptible, exposed (latent), infective, asymptomatic, hospitalized (mild, moderate, and severe cases), recovered, and dead categories of the population, respectively.

and speed of transmission from one compartment to other is based on the transmission rates denoted by arrows. The symbol α represents the transmission rate from susceptible humans to exposed humans. The parameter β denotes the transmission rate from exposed humans to infected/asymptomatic human population. The symbols γ_1 and γ_2 depict the recovery rate from infected and hospitalization, respectively. Further, the parameter δ_1 and δ_2 are death rates from asymptomatic and hospitalization compartment individuals. The symbol σ represents the rate at which infected humans are hospitalized. The range and baseline value of the above parameters used in simulation has been listed in Table 10.1. The resulting mathematical framework of the model is presented by a system of nonlinear coupled ordinary differential equations (10.1a–10.1g). The model is formulated under the following assumptions:

- The human population is sufficiently large so that randomness between individuals can be neglected; therefore, each individual has equal probability to acquire the infection.
- Natural birth and death has been neglected, therefore, the total population always remains constants that is $N = S + E + A + I + J + R + D$.
- We consider that there is no transmission of infection from human to animals and vice versa, that is, infection transmits from humans-to-humans only.

$$\frac{dS}{dt} = -\alpha S \frac{I + J + qA}{N} \tag{10.1a}$$

$$\frac{dE}{dt} = \alpha S \frac{I + J + qA}{N} - \beta E \tag{10.1b}$$

$$\frac{dA}{dt} = (1 - \psi)\beta E - (\gamma_1 + \delta_1)A \tag{10.1c}$$

$$\frac{dI}{dt} = \psi\beta E - (\sigma + \gamma_1)I \tag{10.1d}$$

TABLE 10.1
Parameters with Definitions and Values with Their Range

Symbol	Description	Range	Baseline Value	References
P	Transmission probability	[0,1]	0.33	Assumed
C	Average contact rate	[0,1,5]	1.37	Estimated
B	Infection rate from exposed individuals	$\left[\frac{1}{7},\frac{1}{5}\right]$ days	$\frac{1}{53}$	[12]
γ_1	Recovery rate from infection (in symptomatic or asymptomatic population)	$\left[\frac{1}{14},\frac{1}{7}\right]$ days	$\frac{1}{10}$	[13]
γ_2	Recovery rate form hospitalization/isolation	$\left[\frac{1}{14},\frac{1}{7}\right]$ days	$\frac{1}{10}$	[13]
δ_1	Death rate from asymptomatic population	0%–3%	0.0002	Assumed
δ_2	Death rate from hospitalization/isolation	0%–3%	0.002	Assumed
ψ	Fraction of latent individual into infected population	[0,1]	0.995	Assumed
σ	Hospitalization rate from infected population	[0,1]	0.95	Assumed
q	Reduce transmissibility factor	[0,1]	0.01	Assumed

$$\frac{dJ}{dt} = \sigma I - (\gamma_2 + \delta_2)J \tag{10.1e}$$

$$\frac{dR}{dt} = \gamma_1(A + I) + \gamma_2 J \tag{10.1f}$$

$$\frac{dD}{dt} = \delta_1 A + \delta_2 J \tag{10.1g}$$

The transmission rate α is the product of *transmission probability (p)* and *average contact rate (c)*, that is, $\alpha = p \times c$. The proportion ψ represents the fraction of latent individuals who move to the infected class of individuals. Whereas, $1-\psi$ is the fraction of latent individuals who move to asymptomatic infected class of individuals. Again, the parameter q denotes the reduced transmissibility factor due to the asymptomatic individuals in the infected population of humans.

10.2.2 Basic Reproduction Number

Basic reproduction number (R_0) represents average number of infected individuals by single infected individual when the entire population is treated as susceptible [14]. The basic reproduction number provides knowledge about the progression characteristic of the infection in a given population. This threshold parameter indicates that infection will spread in the population only when the value of R_0 is greater than 1, otherwise infection will not be able to spread in the given population. To calculate the basic reproduction number, we consider the equilibrium of the model such that disease is absent in the population. For this purpose, we observe that at infection-free stage $E = A = I = R = 0$, which implies that $S = N$. Then the above system of equation can be reduced into the following linearized subsystem of equations:

$$\frac{dE}{dt} = \alpha \frac{I + J + qA}{N} - \beta E \tag{10.2a}$$

$$\frac{dA}{dt} = (1 - \psi)\beta E - (\gamma_1 + \delta_1)A \tag{10.2b}$$

$$\frac{dI}{dt} = \psi\beta E - (\sigma + \gamma_1)I \tag{10.2c}$$

$$\frac{dJ}{dt} = \sigma I - (\gamma_2 + \delta_2)J \tag{10.2d}$$

The above system of equations can be written in the matrix form as follows:

$$\frac{d\mathbf{X}}{dt} = (\mathbf{T} + \Sigma)\mathbf{X}$$

where $\mathbf{X} = \begin{bmatrix} E \\ A \\ I \\ J \end{bmatrix}, \mathbf{T} = \begin{bmatrix} 0 & \alpha q & \alpha & \alpha \\ 0 & 0 & 0 & 0 \\ 0 & 0 & 0 & 0 \\ 0 & 0 & 0 & 0 \end{bmatrix}$

and $\Sigma = \begin{bmatrix} -\beta & 0 & 0 & 0 \\ (1-\psi)\beta & -(\gamma_1 + \delta_1) & 0 & 0 \\ \psi\beta & 0 & -(\sigma + \gamma_1) & 0 \\ 0 & 0 & \sigma & -(\gamma_2 + \delta_2) \end{bmatrix}.$

Then, we can compute the next-generation matrix:

$$\mathbf{G} = -\mathbf{T}\Sigma^{-1}$$

$$= \begin{bmatrix} \frac{\alpha\psi}{\gamma_1+\sigma} + \frac{\alpha q(1-\psi)}{\gamma_1+\delta_1} + \frac{\alpha\sigma\psi}{(\gamma_1+\delta_1)(\gamma_2+\delta_2)} & \frac{\alpha q}{\gamma_1+\delta_1} & \frac{\alpha}{\gamma_1+\delta_1} + \frac{\alpha\sigma}{(\gamma_2+\delta_2)(\gamma_1+\sigma)} & \frac{\alpha}{\gamma_2+\delta_2} \\ 0 & 0 & 0 & 0 \\ 0 & 0 & 0 & 0 \\ 0 & 0 & 0 & 0 \end{bmatrix}$$

The basic reproduction number of this modified model is given by spectral radius of matrix $\mathbf{G}$ and is computed as:

$$R_0 = \frac{\alpha\psi}{\gamma_1+\sigma} + \frac{\alpha q(1-\psi)}{\gamma_1+\delta_1} + \frac{\alpha\sigma\psi}{(\gamma_1+\delta_1)(\gamma_2+\delta_2)} = \frac{pc\psi}{\gamma_1+\sigma} + \frac{pcq(1-\psi)}{\gamma_1+\delta_1} + \frac{pc\sigma\psi}{(\gamma_1+\delta_1)(\gamma_2+\delta_2)} \tag{10.3}$$

10.3 SENSITIVITY ANALYSIS OF R_0

The simple model for disease transmission can be solved by analytic approach. But, complex model cannot be solved analytically and therefore numerical methods are used to investigate the disease dynamics through a given model. The model output usually alters as the value of model input parameters change. Sensitivity provides the measure of the variation in model output with respect to change in input parameters. There are different ways in which sensitivity can be defined [15–17]. Sensitivity can be broadly classified into two categories: local sensitivity and global sensitivity. Local sensitivity is related to the partial derivatives of output variable with respect to the input parameters and can be treated as the gradients around the multidimensional reference parameter space [18]. On the other hand, global sensitivity is associated with the overall effect of the input parameters on the model outcome when input parameters vary in a large range. Normalized sensitivity index (elasticity index) is a case of local sensitivity in which model output y and input parameters x are associated together by normalized derivative and defined as:

$$\gamma_x^y = \frac{\partial y}{\partial x} \times \frac{x}{y} \tag{10.4}$$

The basic reproduction number (R_0) is an important threshold quantity which infers about the progression of disease in a given population. It clearly depicts that a disease will spread in a given population if the value of R_0 is greater than unity, otherwise disease will die out in the population. Therefore, it is quite informative to know which parameters influence the R_0 and to what extent. For this reason, sensitivity for R_0 is useful to determine the influenced parameters for the proposed model. We can find an expression for elasticity index of R_0 with respect to the parameter θ based on the formula given in Equation (10.4) below:

$$\gamma_\theta^{R_0} = \frac{\partial R_0}{\partial \theta} \times \frac{\theta}{R_0} \tag{10.5}$$

10.4 RESULTS AND DISCUSSION

We formulated a mathematical model based on SEAIJRD structure as shown in Figure 10.1. We applied the formulated model on the COVID-19 data of ongoing outbreak in India. The used data were collected from the website of Ministry of Health and Family Welfare [19]. The starting date of outbreak is assumed as March 14, 2020.

The model simulation for the different state variables is shown in Figure 10.2. The simulation has been done for the time period March 14, 2020, to September 14, 2020.

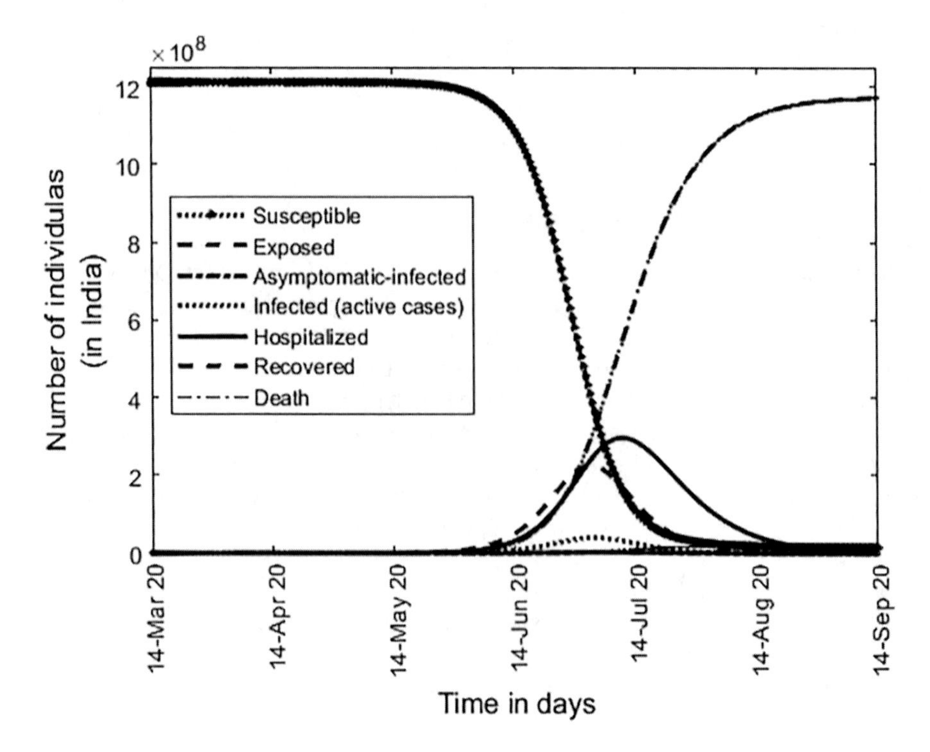

FIGURE 10.2 Model simulation at parameter values: $p = 0.33$, $c = 1.37$, $\beta = 1/5.3$, $\gamma_1 = \gamma_2 = 1/10$, $\delta_1 = 0.0002$, $\delta_2 = 0.002$, $\psi = 0.995$, $\sigma = 0.95$, and $q = 0.01$.

The model has been simulated with the initial conditions for the state variables listed in Table 10.2. The value of parameters during the simulation are: $p=0.33$, $c=1.37$, $\beta=1/5.3$, $\gamma_1=\gamma_2=1/10$, $\delta_1=0.0002$, $\delta_2=0.002$, $\psi=0.995$, $\sigma=0.95$, and $q=0.01$.

The number of actual cases and estimated cases of infected humans for COVID-19 in India are plotted in Figure 10.3. Figure 10.3 shows the best-fit adjusted with the tuning of unknown parameters. The accuracy for this fit is measured by root mean square error (RMSE), and the calculated value of RMSE is equal to 1.40 for

TABLE 10.2
State Variables with Definitions and Initial Values Used in the Simulation

Symbol	Description	Initial Value
$S(t)$	Susceptible humans in the given population	1210854977
$E(t)$	Exposed humans in the given population	70
$A(t)$	Asymptomatic infected humans in the given population	0
$I(t)$	Infected humans in the given population	70
$J(t)$	Hospitalized (mild, moderate, and severe cases) humans in the given population	0
$R(t)$	Recovered humans from infection in the given population	3
$D(t)$	Death due to infection in the given population	0

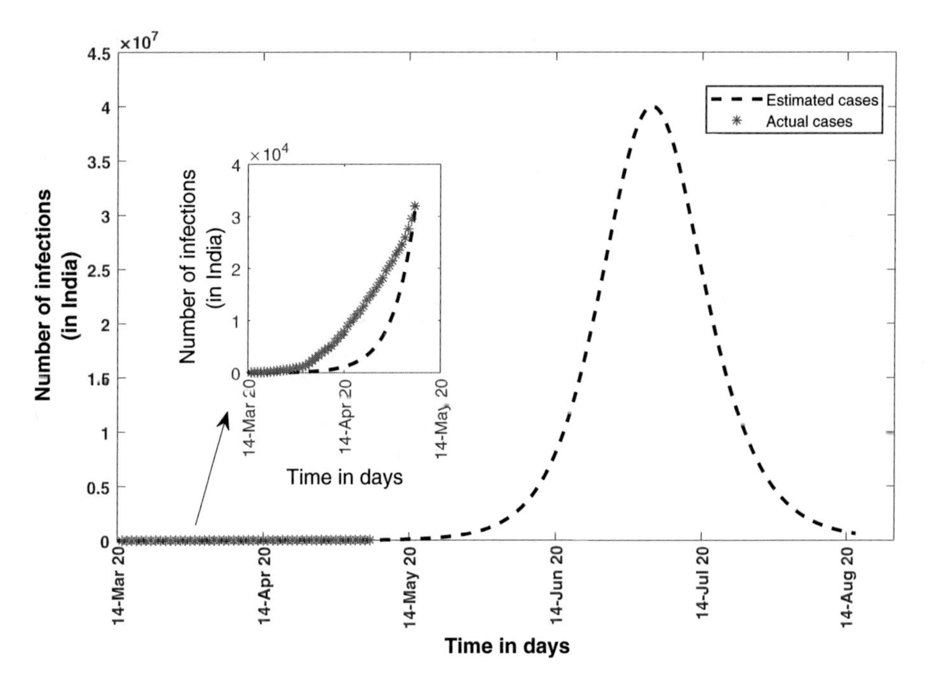

FIGURE 10.3 Estimated and actual cases of COVID-19 in India at parameter values: $p=0.33$, $c=1.37$, $\beta=1/5.3$, $\gamma_1=\gamma_2=1/10$, $\delta_1=0.0002$, $\delta_2=0.002$, $\psi=0.995$, $\sigma=0.95$, and $q=0.01$.

normalized data. This figure suggests that peak of outbreak occurs near mid-June 2020. Further, the analysis depicts that maximum level of active cases for COVID-19 in India will reach approximately 4×10^7 without any intervention program/strategy at the given values of parameters. However, it is not feasible to attain this peak with certain probability as government takes steps on different time interval to cease the transmission of COVID-19. The inset figure indicates that in initial phase of disease transmission, actual cases are greater than estimated cases, which may be the resultant of the occurrence of some hidden clusters in different communities. Furthermore, this figure depicts that transmission of COVID-19 in India will be at minimum level after August 14, 2020.

In the transmission of COVID-19, social distancing plays a major role. The average contact rate (c) is a measure of such social distancing in the formulated model. The best-fitted curve for infected individuals are given in Figure 10.3, which was estimated for the average contact rate, $c = 1.37$. The transmission dynamics of COVID-19 for different values of average contact rate (c) has been shown in Figure 10.4. The figure suggests that effect of average contact rate on the COVID-19 transmission in India is much significant. In this figure, the curves are plotted for $c = 1.37$, 1.2, 1.1, 1.0, 0.9, and 0.8, and these are shown in different shades and corresponding value of basic reproduction number (R_0) is given in Table 10.3. We assumed the hypothetical maximum limit of medical healthcare support facility to be approximately 2×10^7 in India. From this figure, it is obvious that infected individuals are much more (approximately double) compared to the maximum limit of healthcare system at average contact rate $c = 1.37$. Furthermore, the figure depicts that if contact rate reduces to 0.8, then infected individuals peak does not exceed the maximum limit of heathcare

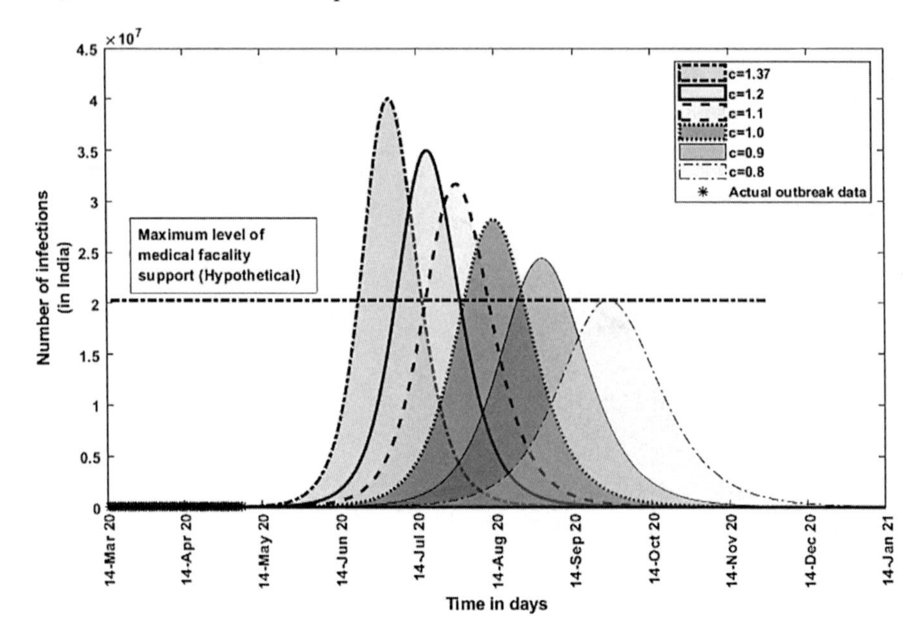

FIGURE 10.4 Effect of average contact rate (c) at parameter values: $p = 0.33$, $\beta = 1/5.3$, $\gamma_1 = \gamma_2 = 1/10$, $\delta_1 = 0.0002$, $\delta_2 = 0.002$, $\psi = 0.995$, $\sigma = 0.95$, and $q = 0.01$.

TABLE 10.3
Value of Basic Reproduction Number (R_0) for Different Values of Average Contact Rate (c)

Average Contact Rate (c)	Value of R_0
1.37	4.4188
1.20	3.8705
1.1	3.5480
1.0	3.2254
0.9	2.9029
0.8	2.5803

support and the peak drops to 2×10^7. It is also observed that peak shifts to the right which would provide time to prepare the management strategy for the government. In Indian scenario, this reduced peak shifts to mid-September 2020. Therefore, average contact rate might be a crucial parameter to examine the disease transmission characteristics in the given population.

The phase-plane curves between susceptible and infected humans are shown in Figure 10.5. In this figure, three phase-plane curves are plotted with respect to three different initial conditions for state variables. The initial condition and scale for infected and susceptible population are taken in proportion with maximum value as 1. The figure depicts that different inflation points (represents peak of infection) exit, but

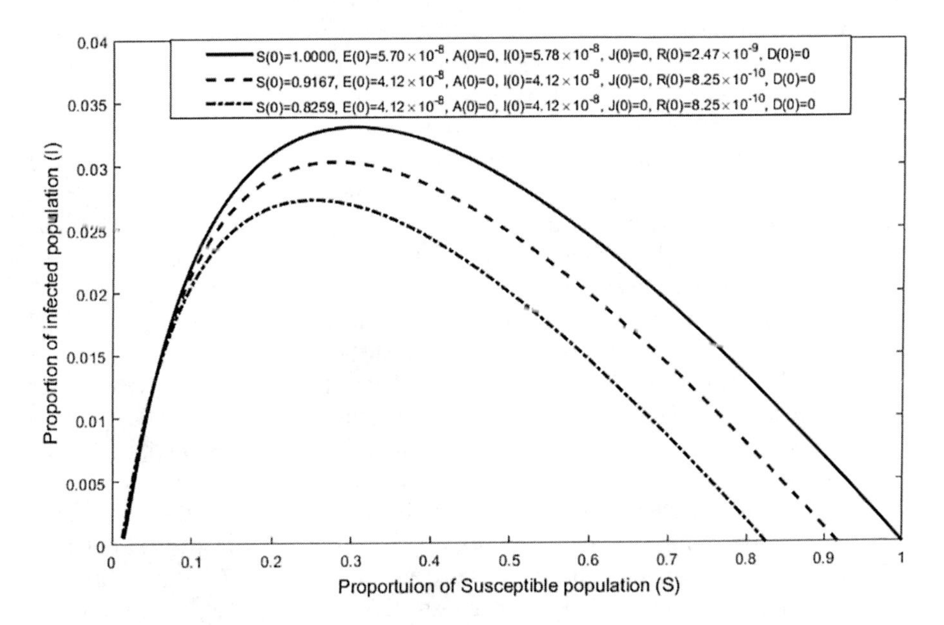

FIGURE 10.5 Phase-plane analysis between infected and susceptible humans for different initial conditions.

TABLE 10.4
Sensitivity Elasticity Expression and Its Value for Basic Reproduction Number R_0 with Respect to Involved Parameters ($p=0.33$, $c=1.37$, $\gamma_1=0.1$, $\gamma_2=0.1$, $\delta_1=0.0002$, $\delta_2=0.002$, $\psi=0.995$, $\sigma=0.95$, $q=0.01$)

Parameter	Elasticity Index Value
p	1
c	1
γ_1	−0.09528418515
γ_2	−0.8852899226
δ_1	−0.000000101904423
δ_2	−0.01770579845
ψ	0.9897891768
σ	−0.00171999185
q	0.0000510541159

finally all curves show same type of dynamics for different initial conditions. Further, it can be seen that susceptible population never approaches 0; however, initially the entire population is treated as susceptible. This behavior of phase-plane curves indicates that whole population will never become infected, that is, some individuals remain without infection in the population for any value of parameters and initial conditions.

The local sensitivity for R_0 is done and related normalized sensitivity index (elasticity index) of R_0 with respect to each parameters involved in the expression of R_0 is given in Table 10.4. The global sensitivity for R_0 is also carried out and the resulting curves are shown in Figure 10.6. Figure 10.6a,b,g depicts that R_0 is an increasing function of transmission probability (p), average contact rate (c), and fraction of latent individuals (ψ). Again, R_0 is a decreasing function with respect to the parameters recovery rate of infected individuals (γ_1), recovery rate of hospitalized individuals (γ_2), and death rate in hospitalized individuals (δ_2). However, R_0 is almost constant for the parameters death rate from hospitalization/isolation (δ_1), hospitalization rate from infected population (σ), and reduced transmissibility factor (q). These curves also validate the result of elasticity index given in Table 10.4 under local sensitivity of R_0. Figure 10.6a,b,g shows that R_0 is highly sensitive for the parameters p, c, and ψ. The value of R_0 can be reduced below the critical threshold value $R_0 = 1$ through these parameters. Again, Figure 10.6b also depicts that the value of R_0 would below 1 when the value of average contact rate is less than 0.4 at given parameter values. This analysis may be treated as a crucial knowledge to design or implement the preventing strategy for COVID-19 in the given population. The effect of other remaining parameters (except p, c, and ψ) on R_0 is not significant in such manner that transmission dynamics of COVID-19 can stop as the value of R_0 does not reduce below 1 by tuning these parameters. The combined effect of the transmission probability (p) and average contact rate (c) on the basic reproduction number (R_0) has been shown in Figure 10.7.

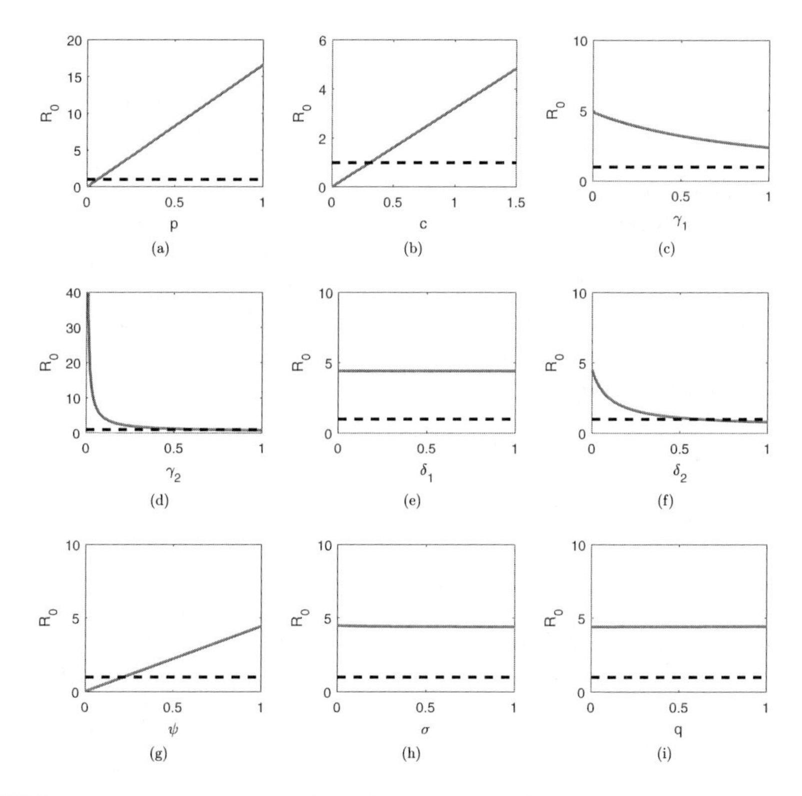

FIGURE 10.6 Global sensitivity of R_0 with respect to different parameters involved in its expression. (a) R_0 vs transmission probability (p); (b) R_0 vs average contact rate (c); (c) R_0 vs recovery rate from infection in symptomatic or asymptomatic population (γ_1); (d) R_0 vs recovery rate form hospitalization/isolation (γ_2); (e) R_0 vs death rate from asymptomatic population (δ_1); (f) R_0 vs death rate from hospitalization/isolation (δ_2); (g) R_0 vs fraction of latent individual into infected population (ψ); (h) R_0 vs hospitalization rate from infected population (σ); (i) R_0 vs reduce transmissibility factor (q).

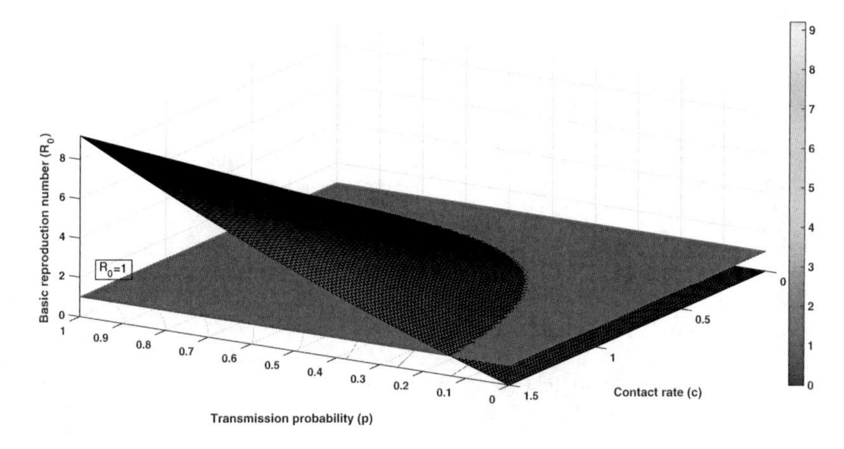

FIGURE 10.7 Sensitivity of R_0 with respect to transmission probability (p) and average contact rate.

10.5 CONCLUSION

COVID-19 has become a serious health threat to the Indian population. Therefore, mitigation strategies are the only solution for invading this pandemic. It is highly recommended to follow the discipline of mitigation strategies. This research work basically proposed a compartmental mathematical model based on the social distancing, quarantine, and isolation. The model estimated the average contact rate for India on the basis of ongoing outbreak data. The average contact rate of individuals and social distancing are highly correlated. Thus, model predicted infection scenario on different contact rate or rate of social distancing discipline following Indian population. The model estimated 1.37 as the current average contact rate. The model suggested maintaining social distancing to invade the COVID-19 pandemic. It is assumed that human population is homogeneously distributed in the model. The involvement of heterogeneity of interaction in the human population exists. The incorporation of this heterogeneity in the model may provide a better prediction about the dynamics of compartmental population.

ACKNOWLEDGMENT

The authors thank the Science and Engineering Research Board, Department of Science and Technology, Govt. of India (under the grant no. EEQ/2016/000509), for their support.

REFERENCES

1. F. Brauer, "Mathematical epidemiology—Past, present, and future," *Infectious Disease Modelling*, vol. 2, no. 2, pp. 113–127, 2017.
2. J. N. Hays, *Epidemics and pandemics–Their impacts on human history*. Santa Barbara, CA, Abc-clio, 2005.
3. H. W. Hethcote, "The mathematics of infectious diseases," *SIAM Review*, vol. 42, no. 4, pp. 599–653, 2000.
4. Website—World Health Organization (WHO). https://www.who.int/. Accessed April 1, 2020.
5. B. Ivorra, M. Ferrández, M. Vela-Pérez, and A. Ramos, "Mathematical modeling of the spread of the coronavirus disease 2019 (covid-19) considering its particular characteristics. The case of China," Tech. Rep., Technical Report, MOMAT, 03 2020. https://doi-org.usm.idm.oclc.org, 2020.
6. J. Cao, X. Jiang, B. Zhao, "Mathematical modeling and epidemic prediction of covid-19 and its significance to epidemic prevention and control measures," *Journal of Biomedical Research and Innovation*, vol. 1, no. 1, p. 103, 2020.
7. A. J. Kucharski, T. W. Russell, C. Diamond, Y. Liu, J. Edmunds, S. Funk, R. M. Eggo, F. Sun, M. Jit, J. D. Munday, et al., "Early dynamics of transmission and control of covid-19—Ā mathematical modelling study," *The Lancet Infectious Diseases*, vol. 20, no. 5, pp. 553–558, 2020.
8. A. S. Shaikh, I. N. Shaikh, and K. S. Nisar, "A mathematical model of covid-19 using fractional derivative—Outbreak in India with dynamics of transmission and control," vol. 2020, no. 373, 2020.

9. M. A. Khan and A. Atangana, "Modeling the dynamics of novel coronavirus (2019-ncov) with fractional derivative," *Alexandria Engineering Journal*, vol. 59, no. 4, pp. 2379–2389, 2020.
10. "European centre for disease prevention and control. Outbreak of novel coronavirus disease 2019 (covid-19)—increased transmission globally—fifth update, 2 march 2020," ECDC, Stockholm, 2020.
11. "Coronavirus disease 2019 (covid-19) pandemic—Increased transmission in the EU/EEA and the UK—seventh update, 25 march 2020," ECDC, Stockholm, 2020.
12. Q. Li, X. Guan, P. Wu, X. Wang, L. Zhou, Y. Tong, R. Ren, K. S. Leung, E. H. Lau, J. Y. Wong, et al., "Early transmission dynamics in Wuhan, China, of novel coronavirus–infected pneumonia," *New England Journal of Medicine*, vol. 382, no. 13, pp. 1199–1207, 2020.
13. Y. Yang, Q. Lu, M. Liu, Y. Wang, A. Zhang, N. Jalali, N. Dean, I. Longini, M. E. Halloran, B. Xu, et al., "Epidemiological and clinical features of the 2019 novel coronavirus outbreak in China," *medRxiv*, 2020. doi: 10.1101/2020.02.10.20021675
14. O. Diekmann, J. Heesterbeek, and M. G. Roberts, "The construction of next-generation matrices for compartmental epidemic models," *Journal of the Royal Society Interface*, vol. 7, no. 47, pp. 873–885, 2010.
15. J. Cariboni, D. Gatelli, R. Liska, and A. Saltelli, "The role of sensitivity analysis in ecological modelling," *Ecological Modelling*, vol. 203, no. 1–2, pp. 167–182, 2007.
16. M. A. Sanchez and S. M. Blower, "Uncertainty and sensitivity analysis of the basic reproductive rate—Tuberculosis as an example," *American Journal of Epidemiology*, vol. 145, no. 12, pp. 1127–1137, 1997.
17. S. M. Blower and H. Dowlatabadi, "Sensitivity and uncertainty analysis of complex models of disease transmission—An HIV model, as an example," *International Statistical Review/Revue Internationale de Statistique*, vol. 62, no. 2, pp. 229–243, 1994.
18. A. Raue, C. Kreutz, T. Maiwald, U. Klingmüller, and J. Timmer, "Addressing parameter identifiability by model-based experimentation," *IET Systems Biology*, vol. 5, no. 2, pp. 120–130, 2011.
19. Website—Ministry of Health and Family Welfare (GoI). https://www.mohfw.gov.in/. Accessed May 10, 2020.

Index